Messy Eating

Messy Eating

Conversations on Animals as Food

Samantha King, R. Scott Carey,
Isabel Macquarrie, Victoria N. Millious,
and Elaine M. Power,
Editors

FORDHAM UNIVERSITY PRESS
New York 2019

Fordham University Press also publishes its books in a variety of electronic formats. Some content that appears in print may not be available in electronic books.

Visit us online at www.fordhampress.com.

Library of Congress Cataloging-in-Publication Data available online at https://catalog.loc.gov.

Printed in the United States of America

21 20 19   5 4 3 2 1

First edition

CONTENTS

Messy Eating

INTRODUCTION

Messy Eating

Many of the conversations featured in this collection of interviews about theory, politics, and animals as food opened with the participants asking *us* a series of questions. Some interviewees were wary of centering their personal dietary histories in an academic forum and wanted to understand what we hoped to achieve by asking about intimate and everyday matters such as meal preparation. Others had agreed to speak to us without hesitation but were curious about the conceptual origins of the work. Like the subject matter we aimed to explore, our answers to these queries were messy, and we often found ourselves sharing different versions of how the project came to be.

Most frequently, we cited a Jeffrey Williams interview with Donna Haraway in which the world-renowned scholar of "significant otherness" is asked if she renounces eating meat.[1] In an intensely conflicted reply, Haraway discusses her desire for a world in which there is a place for agricultural animals, on the one hand, and her inability to articulate an adequate response to the convictions of her vegan friends, on the other. We explained to our interviewees that this dialogue got us thinking about the promise of posing similar questions to other scholars who have devoted their professional lives to theorizing multispecies relationships, but often not theorizing the consumption of animals for food. On other occasions, our genealogy of the project began with a story about a student essay on the "vegan killjoy," or an exchange about Jonathan Safran Foer's *Eating Animals* that occurred at a campus pub (or in a classroom, or at a departmental conference; we have differing memories).[2] We also drew on day-to-day conversations—about our vegetarian "conversion stories," our domestic dietary practices, or our pets (whom we now refer to as "companion animals").

In these exchanges, contributors frequently revealed that they had longed for an opportunity to think through the connections between their theoretical orientations and their eating habits. They seemed to share our sense that while the posthumanist, multispecies, and animal studies literatures with which they themselves were engaged had offered rich insights into a wide variety of interspecies entanglements, these approaches had not quite fulfilled their potential to address the primary site through which such interactions unfold in the contemporary world: the human consumption of nonhuman animals for food. Nor, in the view of many of our participants, had these bodies of work grappled adequately with the messy, persistent dynamics of race, sex, ability, and colonial capitalism that shape species differentiation and commingling in the context of food. The interviews would, we hoped, allow participants to contemplate how academic work on multispecies living that is attuned to such hierarchies might inform contentious public debates about pressing issues ranging from food sovereignty and sustainability, to the exploitation and suffering of human and animal farm laborers. Moreover, we anticipated that by asking scholars to describe how they navigate concerns about food in their daily lives, we might offer an accessible and grounded entry point into such questions.

There is of course a considerable scholarly and popular literature on "happy meat," "ethical eating," and the potential of small-scale agriculture and mindful consumption to address the challenges of contemporary food systems.[3] Work in this area includes the veritable explosion of investigative journalism and memoir documenting what Chad Lavin calls "adventures in immediate food."[4] In such texts, authors share their desire to develop a more proximate, authentic, and informed relationship to the content of their diet. They do not promote vegetarianism or veganism *per se* but instead work to redefine particular approaches to meat eating as ethical—an approach that Kelly Struthers Montford pointedly challenges in her response to the interviews featured in this book.[5] A different set of literatures—writings on meat, vegetarianism, and animal rights theory—provoke in the reader what Jon Mooallem calls "digestive dissonance," or physical repulsion toward the consumption of animal flesh through graphic stories about the violence and unsanitariness of the factory farm and the slaughterhouse.[6] Despite their divergences, taken together these analyses highlight the disastrous effects of industrial agricultural production on food quality, animal welfare, human health, local economies, environmental pollution, and climate change. But they also tend to assume the universality and boundedness of species categories and rarely attend to the shared, affective, and embodied character of human–animal entanglements, or to the political promise of thinking and acting "with and alongside the nonhuman world."[7] Moreover, such works commonly imagine that transformation will occur through a conscious consumerism dependent on bourgeois and gendered systems of food provisioning, or through gradual changes to law and policy grounded in an ethical framework that accords priority to animals who are thought to most closely resemble humans.[8]

In short, literature on the ethics and politics of food and literature on the ethics and politics of human–animal relationships have infrequently converged—a divide that might be said to reproduce humanistic and dualistic thinking about animals as objects that humans eat versus animals as subjects with whom humans relate. *Messy Eating* represents an initial step toward bridging this divide. Speaking with Canada- and U.S.-based scholars at a variety of career stages who do critical interdisciplinary work related to animals, we conducted interviews that explore how postcolonial, Indigenous, black, queer, trans, feminist, disability, continental, phenomenological, posthumanist, and multispecies theories shape approaches to consuming animals as food; how researchers weave their knowledge practices with their ethical and political practices as they conceptualize and in some cases undertake the ingestion of animals;[9] and how the daily, intimate, sensual, and visceral practices of food purchase, preparation, and consumption might enable or constrain thinking about multispecies relationships. We chose to *speak* with researchers because we hoped to highlight the "push and pull" of thought not captured in traditional book or article genres.[10]

Beyond Animal Rights

Early on in our process, we made a decision to feature scholars who operate largely outside liberal and universalist discourses of suffering and rights associated with a traditional animal liberation framework.[11] We did so for three main reasons: First, we wished to move beyond well-rehearsed debates about the comparative sentience of humans and animals and the use of human–animal difference or similarity as the basis for ethical decision making. As Chad Lavin argues, both permissive and restrictive approaches to eating animals "trade in a notion of 'the human' that has come under increasing fire in recent years from various angles."[12] From the shifting science of speciation, to innovations in cross-species engineering, and from Indigenous ontologies of relational personhood, to new humanist, post-colonial, poststructuralist, feminist, queer, and disability studies perspectives that challenge enlightenment notions of an autonomous, biocentric subject, demarcations between humans and animals are increasingly understood as historically produced, permeable, fluid, and contingent. Within these frameworks, the human emerges as just one life form among many.

While the scholars featured in this collection work across a range of theoretical traditions and diverge considerably in their food practices and politics, they have in common a tendency to shy away from approaches that extend the category of the human outward to include (certain) animal others. Instead, they tend to understand and inhabit the world from what interviewee Matthew Calarco calls "perspectives other than those of the classical human subject," often emphasizing "how human beings might find themselves within or alongside animal life in surprising ways."[13] Some contributors are less eager than others to discard the perspective or category of the human, revealing themselves to be oriented toward what Katherine McKittrick, following Sylvia Wynter, describes as "undoing and unsettling—not replacing or occupying—Western conceptions of what it means to be human."[14]

The eurocentric normativity of much research in posthumanist, multispecies, and critical animal studies constitutes the second reason for our moving away from traditional frameworks. We have been motivated by critiques of such work and are committed to profiling scholars whose work challenges the epistemic politics and exclusions of these fields.[15] Whereas ethical vegan critiques of posthumanism have focused on its romanticization of hybridity and its equivocation with regard to violence against animals, our skepticism emerges from a slightly different set of concerns:[16] Like Alexander Weheliye, we are wary of utopic visions that suggest "We have now entered a stage in human development where all subjects have been granted equal access to western humanity and this is, indeed, what we all want to overcome."[17] And while we value posthumanism's attention to the liveliness and agency of nonhuman matter, we share interviewee Neel

Ahuja's observation that posthumanism tends to "project upon an outside, the nonhuman, the possibility of resistance to anthropocentrism" and thus obscures the social and economic systems that reproduce injustice and inequality.[18] Here, Ahuja's perspective overlaps with fellow interviewee Kari Weil's concern that posthumanist theorizing can sometimes "absolve humans of our responsibility for the kinds of agency we do have in the world, agency that has often been destructive." Billy-Ray Belcourt, a respondent to the collection, approaches the limitations of the related field of critical animal studies (CAS) from yet another angle, drawing attention to CAS's failure to center an analysis of the settler colonial state, the specificity of which gets subsumed in the intersectional approach favored by scholars working in this area.[19]

While the scholarship of the authors featured here tends to focus principally on human–animal relations, rather than on food or meat, our efforts to challenge existing frameworks are inevitably informed by work that brings questions of race and ethnicity to bear on the messiness of food and "eating cultures" in the North American context.[20] Resonating with the work of scholars such as Kyla Wazana Tompkins, Anita Mannur, and Martin F. Manalansan IV, our interviews highlight how food and eating emerge as key sites through which racial and colonial subjectivities, hierarchies, and anxieties are produced, experienced, and interpreted.[21] These scholars' shared attentiveness to the corporeal, visceral, and affective dimensions of consumption, and to the "circuitous" routes "between the senses, memory, social location, and history," is also echoed in the embodied experiences, contradictions, and complexities that our conversations capture.[22]

Questioning white universalism and racial exclusion has not been a central concern for every scholar featured in this collection. Indeed, the work of some of our interviewees has been subject to critique on these very grounds, and we expanded our initial focus beyond posthumanist and critical animal studies precisely because we were finding we could not get at the kinds of issues we wanted to address—around race and colonialism especially—by speaking only with people working within those fields narrowly defined. Thus, *Messy Eating* might best be considered a snapshot in a time of a diffuse and wide-ranging conversation that is moving along a trajectory shaped by the editors' own learning. While the ultimate lineup of scholars is diverse with regard to how each of the participants traces the particular configurations of power imbricated in multispecies and dietary subjectivities and systems, taken as a whole, the collection works against the tendencies of both posthumanism and critical animal studies to disregard analyses of race, sex, gender, and colonial capitalism. Instead, it centers the insights of scholars who refuse to privilege the human, *or* the animal, as the primary subject or category of analysis.

While many of our interviewees are indebted to animal rights literature, especially its feminist traditions, and while there are vegetarians and vegans among our interviewees, our purpose was to speak with scholars whose dietary politics are not made explicit in their work and for whom veganism is not a starting place or end point for theory or politics—our third reason for decentering this literature.[23] We tend to agree with Neel Ahuja, who describes veganism in his interview as an "important yet limited tactic for people in countries where there is an advanced factory-farming system that is industrialized." Veganism, he notes, can be mobilized progressively or regressively, and we are interested in when and how those trajectories unfold.

In highlighting the fraught character of discourse on—and of—veganism, *Messy Eating* may also remind readers of the widely read e-mail debate between queer literary scholars Will Stockton and Karen Tongson that was reprinted on the *Bully Bloggers* website in August 2015.[24] Stockton and Tongson had engaged in a heated exchange on Facebook prompted by the death of Cecil the lion, who had been shot, skinned, and beheaded by Walter Palmer, an American dentist and big-game hunter, in Matabeleland North, Zimbabwe, the previous month. In response to widespread outrage about Cecil's death, Stockton, a committed vegan, had asked his Facebook community about the difference between killing Cecil and killing "any cow, chicken, pig, or fish." Tongson argued, in response, that there is an important difference between the elitism and colonialism of big-game hunting and the eating of animals as a "deep expression of cultural heritage and belonging."[25]

The discussion that Stockton and Tongson went on to have in the slower environment of e-mail coincides with many of the key questions and themes that emerge in the chapters which follow, including the desirability and practicability of universal codes of ethics in relation to animals and foods. As an editorial team, we are not interested in judging which dietary practices are most or least preferable—indeed, we seek to trouble the bourgeois individualist underpinnings of the very notion of "ethical eating." Rather, our goal has been to capture, through the scholarly interview, the complex and contradictory ways in which theories and practices interact in orientations to food. Our approach is inspired by Alexis Shotwell's suggestion that it is useful to think about "complicity and compromise as a starting point for action."[26] And we share an outlook articulated by Elizabeth Costello, the eponymous protagonist and animal rights advocate of

J. M. Coetzee's novel, who, when asked by an audience member "what she would have people do—shut down factory farms? stop eating meat?" says she has "never been much interested in proscriptions" but in "what lies behind them."[27]

The Interviews

The process through which *Messy Eating* came together was organic and contingent but remained a collective enterprise throughout. We are a multigenerational and multidisciplinary editorial group—members range in age from twenty-six to fifty-eight and among us hold degrees in biochemistry, cultural studies, film, history, kinesiology, nutrition, public health, sociology, and sport studies. Despite our divergent academic communities, we had met only two of our interviewees prior to this work, so in most cases we "cold" e-mailed potential recruits. Once people agreed to participate, we arranged interviews that took place in person, on the phone, or online and were conducted by at least one, but usually two or three, members of the team. We sent scholars a list of possible questions in advance (with a promise that they would have the opportunity to edit their transcript), but we tried to approach our subject matter through free-flowing and exploratory conversation rather than a rigid interview protocol. Some questions were asked of all participants, but we also developed custom questions based on our readings of each scholar's work. Participants often gave us feedback on the interview guide, as well as suggestions for other people to interview. Relatively few scholars turned down our request for a conversation, which meant that we were unable to move very far down our list of potential participants, especially once we started following the suggested leads. The result is an eclectic mix of voices from which emerge some common threads and themes, especially with regard to theoretical influences. One prepublication reviewer noted the number of participants (four that we know of) with close personal or professional connections to Donna Haraway—a circumstance that was unplanned, but also unsurprising given her pervasive influence. In this sense, the book reflects how networks and communities develop and cohere around shared interests and perspectives that are not necessarily knowable in advance.

Going into the conversations, we only occasionally had a sense of a participant's dietary preferences and practices, and our own eating habits came up only on occasion. It turns out that seven of the thirteen scholars whom we interviewed (Neel Ahuja, Matthew Calarco, Lauren Corman, Naisargi Dave, Maneesha Deckha, María Elena García, and Sunaura Taylor) eat a vegan diet on a consistent basis. For the record, we are an inconsistent crew who, if forced to identify today, would count an omnivore, a pescetarian, a vegetarian, a wishy-washy vegetarian/locavore wanna-be, and a vegan-at-home among our numbers. While we recognize the potential that a few more omnivores or an "unrepentant meat eater" would have brought to the text, we can assure readers that there is plenty of generative friction among this veg*n-leaning group.[28]

The conversations that follow cover a vast swath of theoretical, political, and personal terrain. While they unfold with varying degrees of intimacy, they are, without exception, provocative and poignant. Through narratives that are at once theoretical and experiential, these scholars capture the challenges and pleasures of food provisioning, preparation, and consumption, consistently highlighting the multidimensional, complicated character of these practices.

Several contributors offer extended analyses of ways to think about purity and paradox, specifically in relation to veganism. Naisargi Dave condemns "the tyranny of consistency" that prevents people from adopting practices such as veganism because of a belief that they will have to commit completely and uniformly: "People ask: 'Isn't there a contradiction in the fact that you care about animals in a way that coincides with what the Hindu right thinks about cows or animal protection?'" But, she continues, "that contradiction doesn't actually exist between two things but exists in the frame, in our concepts of things, not in the things themselves." For Matthew Calarco, the "logic of contiguity," where one ethical question about consumption "abuts to another one" and generates question after question after question, leads to an understanding that there is no innocence, that "violence against animals pervades and saturates the social sphere" and that there is no way to avoid the messiness of eating. Because, for Calarco, veganism provokes exactly this kind of constant curiosity, it opens up new ways of understanding and experiencing the world—a perspective that stands in contradistinction to stereotypical notions of veganism as rigid and pietistic. In his Coda to the book, Billy-Ray Belcourt discusses killing as care in the context of Inuit relations with seals and other animals. Using the example of Inuit singer Tanya Tagaq's much-publicized rebuke to vegan activist organization PETA (People for the Ethical Treatment of Animals), Belcourt argues for the generative capacities of paradox, thus countering the idea that contradiction presents an impasse to political action: "[W]orlds, especially Indigenous ones, proliferate paradox," he writes.

Discussions of paradoxes pertaining to hunting, killing, and "being prey" emerge repeatedly through the text.[29] For instance, both Kim TallBear and

Matthew Calarco emphasize that human and nonhuman animals are potential meat, but they land in different places—cannibalism versus veganism—as they think through the implications of this idea. Says TallBear: "I actually come to see now how certain peoples practice cannibalism. Because if we live in relation to nonhumans, it also makes sense that we should be able to eat our own if there was nothing so elevated and special about us." Whereas Calarco, who is "moved" by scholars who "drag human beings into the edible sphere . . . back into the flows of life," questions how to eat and be eaten respectfully and wishes to grant animals, like humans, the opportunity to live in the world as more than meat.

Emphasizing that eating is irreducible to *a priori* ethical codes and a politics grounded in surety or simplicity, interviewees move easily past our questions about their relationships to eating meat to consider messiness in relation to eating other beings, such as plants, and to the myriad ways humans relate to animals in their often mutual capacities as teachers, students, workers, sources of clothing, and subjects of academic work. With at least two equestrians and two dog trainers in the mix, as well as participants who are ambivalent about all human relationships with domesticated animals, including "pets," discussions about interspecies companionship recur throughout the text.

The scholars featured in this collection understand the absence of purity in their interactions with animals not as a moral failure but rather as a starting point from which to consider and negotiate ethical relations as they unfold, thus rejecting any notion of an *absolute good.* Cary Wolfe explains that the ethical undecidability fostered in contemporary continental philosophy helps students, in particular, connect with the animal question. He finds that his students tend to shy away from humanist or liberal animal rights philosophy in which there is "a formula in your back pocket that you can just whip out, prescribing what is the right thing to do in any particular situation." Wolfe's students, like most of our interviewees, hold conceptions of ethics that do not equate the ethical with nonviolence, or the unethical with violence, because they understand that it is impossible to stand with moral clarity at one end of a dichotomous opposition, absolved of guilt, in our "messy" contemporary political landscape. Moreover, these scholars do not seem to think a morally simple (or simplistic) position is even desirable. They conceive of messiness as *productive*, bringing fresh perspectives and new models for "living well."

That said, the interviewees' common suspicion toward the apparent simplicity of foods labeled in some way as "ethical" (e.g., organic, free-range, etc.) means that they rarely feel good about their consumption practices and frequently reference the skepticism, guilt, and frustration that haunt them as they navigate this aspect of their daily lives. Kim TallBear, who grew up on the South Dakota reservation of the Flandreau Santee Sioux Tribe and later in St. Paul, Minnesota, notes, "[E]ven though I'm this middle-class, conscious person now, I'm sure I ate in a much more sustainable, humane way when I was a kid because we procured so much of our food from our own labor." Kari Weil describes the extensive list of things she "looks for" when making purchases but goes on to say, "I also have little faith that these labels mean very much, but I'm kind of stuck with them." And H. Peter Steeves says, "Unless you're producing your own food, you are supporting systems that are inherently horrible for fostering real community and ethical living."

Steeves is one of several authors who address their ambivalent relationship toward conscious consumerism in the context of contemporary capitalism. He decries the "tricks" that liberalism and capitalism play "to make us think we're free," noting that when it comes to food provisioning, "there are a lot of options, but not *really* very many." Our conversation with Neel Ahuja helps expand this analysis to consider the devastating impact of trade agreements and agricultural subsidies on countries in the Global South. Ahuja notes how "meat eating is experienced as a luxury in most parts of the world," and "how the capacity to be vegan grows with one's economic means," at least in a society like that of the United States, where animal-based foods are "artificially cheap."

Alongside their critiques of capitalism, the interviewees express impatience with the pretensions and privileges associated with foodie cultures; however, they also emphasize the community building and political potential of collective food preparation and appreciation. Sharon Holland cooked for herself and her mother from a young age and captures the love and joy that can be communicated through feeding others: "Around the time of my father's death, I also took solace in the kitchen. To avoid writing my dissertation I got the ginormous copy of Lord Krishna's *Vegetarian Cuisine* and worked my way through Indian cooking. I used to have these big dinners where people would come at nine o'clock thinking that food would be ready and I would be like, 'Oh, you need to soak that for an hour.' And they would eat until two in the morning. I don't know what I was doing, but they were so much fun and people would stay in the house until dawn just eating, playing cards, and talking politics." Harlan Weaver notes that cooking has been central to the development of his intellectual communities from the early days of his undergraduate studies: "I went to college at Wesleyan in Connecticut, which was amazing; I loved it there. I majored in what was then 'Women's Studies' and is now called 'Feminist, Gender, and Sexuality Studies.' And part of growing into that community for me involved collective cooking." Weaver describes how "hanging out in the kitchen" was central to his experience as a doctoral student working with Donna Haraway in the History of Consciousness program at Santa Cruz. "Donna, and pretty much all of Donna's advisees, loved to cook," he notes.

The impact of profound injustices, as well as pleasures, on the development of interviewees' academic orientations and food practices is also evident. Sunaura Taylor describes how growing up in a neighborhood of Tucson, Arizona, that was contaminated by military toxins—the cause of the disability with which she lives—encouraged her to investigate environmental racism and cohabitation and dependency across species. She says, "I just had this sense as a child that my body was the way it was because of how poorly people treat each other and the environment. That was the understanding I had as a young kid, which sort of gave me this identity as an activist. I felt this responsibility as a kid, and I think on one level it also gave me meaning, gave me a way to understand being disabled." María Elena García traces her interest in colonial projects and gourmet food cultures through her family's departure from Peru in the midst of the war between the state and the Shining Path, noting the disproportionate impact of the conflict on Indigenous people: "I have been struck by the fact that animals don't really figure in narratives of political violence, except for a couple of iconic cases. . . . Many of the testimonies of Indigenous Peruvians include the theft or mutilation or killing of their animals as violence they experienced. I want to go back through those testimonies, to track the animals throughout, and to talk with Native and *campesino* activists about this project and approach." Both Naisargi Dave and Neel Ahuja extend their thinking about Hinduism beyond the personal and the familial to address how vegetarianism and the figure of the cow get mobilized in the context of Hindu nationalism and anti-Muslim violence. Regardless of the specific examples through which interviewees discuss their research, they share a commitment to theorizing the relationship of racial and colonial formations to animal oppression not as analogical or competitive, but as co-constitutive. In this regard, the interviewees display what Claire Jean Kim labels "multi-optic vision," or the embrace of an "ethics of mutual avowal" that acknowledges and seeks to actively connect what are often theorized or enacted as separate struggles.[30]

For our participants, multi-optic vision tends to manifest in a rejection of paternalistic attitudes toward animals and a preference for emphasizing the shared vulnerability of species. Sunaura Taylor ties her gradual shift toward veganism to both her recognition that she needed help with the activities of daily living and the development of her scholarly interest in mutual reliability among humans and nonhuman others: "[B]ecoming vegan was perfectly aligned with coming to terms with the fact that I needed help, and that I needed to hire someone to help me. So in that way I felt that the coming into realization of veganism and coming into a political awareness of my own body, and not feeling ashamed that I needed help, was actually about figuring out a practice of interdependence." For Kim TallBear, animals have "their own life trajectories." TallBear does not desire intimate relations with animals for herself, but, like others featured in the collection, she makes the case for thinking relationally and nonhierarchically *with* the more-than-human world. Similarly, Harlan Weaver has written about "becoming-in-kind" with his pitbull terrier, Hayley, as Weaver transitioned from female to male, and Hayley helped ensure his safety in moments of vulnerability.[31] While Hayley helped make Weaver's "gender expression possible," Weaver describes how he became attuned to the ways that both anti-pitbull activists and advocates relate to these dogs through ideas about "black and brown masculinities" and "messed-up racialized language," that he has felt compelled to interrogate.

Despite their shared critique of mainstream animal rights discourses, several contributors describe their exposure to animal rights activism at a young age as key to their politicization, the later development of their scholarly interests, and the transformation of their eating practices. Neel Ahuja and Cary Wolfe both reference being provoked by pamphlets about cruelty: Ahuja became a vegetarian in fifth grade following a classroom debate about vivisection, but several years later, a flyer in a bathroom stall about industrial slaughterhouses prompted him to learn more about the food system and eventually to form a campus animal activism organization. For Wolfe, an encounter as an undergraduate with animal rights materials about biomedical research on primates resulted in an intense sense of "not being able to turn away." Matthew Calarco and Lauren Corman were inspired by books and a video, respectively, and both scholars describe being profoundly disturbed by what they saw and an almost instantaneous change in their approach to eating. Calarco says, "I'll be honest, I was shaken, I was deeply shaken. I was horrified by it. As *soon* as I saw what was going on, and had some sense of what was going on, I switched, *automatically*. I became vegetarian *instantly*. And it wasn't a struggle; it wasn't difficult." Corman describes her viewing of a vivisection similarly: "That was a crushing moment, and it was really terrifying too. There was this monkey who was screaming, and I guess that was the moment of connection between the animals I eat and this particular animal who was clearly suffering. I went vegetarian that night, and then pretty quickly vegan after that."

But transformative new knowledge about animal suffering was not the only pathway to dietary change. Several scholars express skepticism about the transformative capacities of schooling and education, and the interviews make clear that it is often the people and contexts through which knowledge circulates that prove most influential in shaping approaches to food. For several interviewees, both sudden and slowly evolving changes in their dietary preferences were bound up with desire, or with intimate or professional relationships that seemed to inspire and irritate the scholars in equal measure. Naisargi Dave tells us that she changed her eating habits to impress a "hot woman" whom she saw mowing the lawn of her rented house one day. Dave learned from her landlord that the woman was a vegan, and even though the two never actually met, Dave maintains a plant-based diet to this day. Dates and debates with vegans highlighted for Kim TallBear how Indigenous peoples' relationships to animals and the land are erased or even pathologized in purist frameworks: "I was actually dating a vegan until I left Austin for Edmonton. I don't know if all vegans are like this, but I would call him a very typical vegan where he's got this sense that he can stand in this morally pure place because he's a vegan. Yet he shops at Whole Foods, and he buys fruits and vegetables that are being shipped using fossil fuels, and his soy is being produced in the Amazon and that's displacing humans and nonhumans." Harlan Weaver describes dating someone who "was not really into vegetarian stuff" and gradually switching back to meat eating as a result, and Kari Weil speaks of buying meat for her husband, stating that "I don't want my relations with animals to make it look as though I condemn my friends who have other relationships with animals."

These discussions of food in the context of intimate relationships connect to another major theme that emerges from the interviews: the personal dimensions of professional academic life.[32] Across all the conversations, it is clear that work comes home; that meal planning, hosting, and socializing extend into the academic world; that the desires, affects, and bonds that constitute family, friendship, class, and community shape how scholars find research projects and engage politically; and that intellectual pursuits can both enable and constrain such connections. For instance, both Neel Ahuja and H. Peter Steeves discuss food purchases in professional contexts. Whereas Ahuja references his discomfort with academics spending "huge amounts of money on fancy meals," regardless of the type of cuisine in question, Steeves highlights his decision to not spend university money on animal-based foods when entertaining visiting scholars. "[T]hat has been challenging at times," he says, "because the question of hospitality can conflict with my strong moral commitment. . . . [T]here have only been two times when someone became so distressed by the lack of meat that they demanded some. . . . That's an uncomfortable conversation to have because it's hard not to feel superior, or to talk down to someone—both things I just loathe. I'm not much of an outward proselytizer; I just try to be an example."

For some scholars, their current food practices are consistent with household norms, whereas for others a change in their diet, which is often connected to their professional work, has become a source of tension and a site of delicate negotiation with family members. Maneesha Deckha claims that the normalization of vegetarianism in her Hindu childhood home made veganism a fairly uneventful step, whereas Naisargi Dave recalls how her vegetarian mother, whose "neighborhood in Ahmedabad is a very Brahminical one," "was actually more devastated about my being vegan than about my being a lesbian." Neel Ahuja's parents, who "largely ate vegan," were also resistant to his vegetarianism and eventual veganism, as was María Elena García's mother, even though García's mother holds a PETA membership. Cary Wolfe's mother experienced his decision to no longer eat animals as "an attack on her being a good mother," but in responding this way she provided an important lesson: "A kind of light bulb went on; I realized that whenever you are talking about these ethical issues around food and eating, the worst kind of fantasy you can have is that they all make sense."

Maneesha Deckha and Sunaura Taylor both discuss veganism in relation to parenting and to the particular forms of surveillance they experience in their capacity as mothers. Deckha responds to a question about if and how her dietary practices have changed over time with the following: "Yes, I am still a vegan. I can't really say my thoughts have changed over time about that. It's not really a preoccupation for me; I guess these days I'm more preoccupied with the judgment I receive for having my young child be vegan. I've become aware of the sensationalist media stories about vegan mothers being charged with neglecting care and things like that." Taylor's description of her experience helps draw out how scrutiny of mothers unfolds differentially: "On one level, all of the medicalization that happens around being a very visibly disabled woman who was pregnant totally took away from any concern people had about my being vegan. It was like my veganism was so low down on the list of what people were freaked out about that it really did not come up."

This moment in Taylor's interview—when we asked about mothering and food without thinking, as well, about disability—is one of several that continue to reverberate for the editorial team. We embarked on this project with a great deal of excitement, nervousness and naïveté, with little sense of how generative or provocative the interviews would be—of how many "aha moments" we would experience. We share this reflection as a way to acknowledge how grateful we are for the discussions we have had and for what we have learned. An edited volume was only one possibility for how this work would unfold in the early days, so we also appreciate that our interviewees were willing to participate with little sense of where it would lead (or in many cases of who we were). They were generous in their suggestions about whom we might approach for interviews, and whom and what we should read, watch, and think alongside. This engagement undoubtedly improved the breadth and thoughtfulness of our research while it was in progress. Without our interviewees' willingness to speak with us, candidly and openly, this book would not have been possible.

NOTES

1. Jeffrey J. Williams, "Science Stories: An Interview with Donna J. Haraway," *The Minnesota Review* 73–74 (2009–10): 161.

2. Jonathan Safran Foer, *Eating Animals* (London: Penguin, 2009); Richard Twine, "Vegan Killjoys at the Table: Contesting Happiness and Negotiating Relationships with Food Practices," *Societies* 4, no. 4 (2014): 623–629.

3. Brenda L. Beagan, Svetlana Ristovski-Slijepcevic, and Gwen E. Chapman, "People Are Just Becoming More Conscious of How Everything's Connected: 'Ethical' Food Consumption in Two Regions of Canada," *Sociology* 44, no. 4 (2010): 751–769; E. Melanie DuPuis and David Goodman, "Should We Go 'Home' to Eat? Toward a Reflexive Politics of Localism," *Journal of Rural Studies* 21, no. 3 (2005): 359–371; Catherine Friend, *The Compassionate Carnivore: Or, How to Keep Animals Happy, Save Old MacDonald's Farm, Reduce Your Hoofprint and Still Eat Meat* (New York: De Capo Press, 2009); Julie Guthman, "Fast Food/Organic Food: Reflexive Tastes and the Making of 'Yuppie Chow,'" *Social and Cultural Geography* 4, no. 1 (2003): 45–58; Josée Johnston and Michelle Szabo, "Reflexivity and the Whole Foods Market Consumer: The Lived Experience of Shopping for Change," *Agriculture and Human Values* 28, no. 3 (2011): 303–319; Josée Johnston, Michelle Szabo, and Alexandra Rodney, "Good Food, Good People: Understanding the Cultural Repertoire of Ethical Eating," *Journal of Consumer Culture* 11, no. 3 (2011): 293–318; Karyn Pilgrim, "'Happy Cows,' 'Happy Beef': A Critique of the Rationales for Ethical Meat," *Environmental Humanities* 3, no. 1

(2013): 111–127; Safran Foer, *Eating Animals*. ebook (Chapter 6, "Slices of Paradise/Pieces of Shit")

4. Chad Lavin, *Eating Anxiety: The Perils of Food Politics* (Minneapolis: University of Minnesota Press, 2013).

5. Barbara Kingsolver, *Animal, Vegetable, Miracle: A Year of Food Life*, first edition (New York: HarperCollins, 2007); James E. McWilliams, *Just Food: How Locavores Are Endangering the Future of Food and How We Can Truly Eat Responsibly* (New York: Little, Brown, 2010); Michael Pollan, *The Omnivore's Dilemma: A Natural History of Four Meals* (New York: Penguin, 2006); Michael Pollan, *Food Rules: An Eater's Manual* (New York: Penguin, 2009).

6. Jon Mooallem, "Carnivores, Capitalists, and the Meat We Read," *Believer*, 3 (October 2005). http://www.believermag.com/issues/200510/?read=article_mooallem. Work of this sort includes: Timothy Pachirat, *Every Twelve Seconds: Industrialized Slaughter and the Politics of Sight* (New Haven, Conn.: Yale University Press, 2011); Eric Schlosser, *Fast Food Nation* (Boston: Houghton Mifflin Harcourt, 2001); Matthew Scully, *Dominion: The Power of Man, the Suffering of Animals and the Call to Mercy* (St. Martin's Griffin, 2003).

7. Matthew Calarco, "Identity, Difference, Indistinction," *CR: The New Centennial Review*, 11, no. 2 (2011): 45.

8. Ibid.

9. Here we refer to the conscious consumption of animals on the part of humans, not the unavoidable consumption of insects and microbes.

10. Jeffrey J. Williams (ed.), *Critics at Work: Interviews, 1993–2003* (New York: New York University Press, 2004), 2.

11. For prominent representatives of this tradition, see Tom Regan, *The Case for Animal Rights* (Berkeley: University of California Press, 1983); Peter Singer, *Animal Liberation: A New Ethics for Our Treatment of Animals* (New York: Avon, 1975).

12. Lavin, *Eating Anxiety*, 116.

13. Calarco, "Identity, Difference, Indistinction," 45.

14. Katherine McKittrick (ed.), *Sylvia Wynter: On Being Human as Praxis* (Durham, N.C.: Duke University Press, 2015), 2.

15. Neel Ahuja, *Bioinsecurities: Disease Interventions, Empire, and the Government of Species* (Durham, N.C.: Duke University Press, 2016); Billy-Ray Belcourt, "Animal Bodies, Colonial Subjects: (Re)locating Animality in Decolonial Thought," *Societies* 5, no. 1 (2014): 1–11; Breeze Harper, *Sistah Vegan: Black Female Vegans Speak on Food, Identity, Health and Society* (New York: Lantern Books, 2009); Zakiyyah Jackson, "Animal: New Directions in the Theorization of Race and Posthumanism," *Feminist Studies* 39, no. 3 (2013): 669–685;

Claire Jean Kim, *Dangerous Crossings: Race, Species, and Nature in a Multicultural Age* (Cambridge: Cambridge University Press, 2015); Claire Jean Kim and Carla Freccero, "Introduction: A Dialogue," *American Quarterly* 65, no. 3 (2013): 461; Anthony J. Nocella II, John Sorenson, Kim Socha, and Atsuko Matsuoka, "The Emergence of Critical Animal Studies: The Rise of Intersectional Animal Liberation," in *Defining Critical Animal Studies: An Intersectional Social Justice Approach for Liberation* (New York: Peter Lang, 2014); Stephanie Jenkins and Vasile Stanescu, "One Struggle" in *Defining Critical Animal Studies: An Intersectional Social Justice Approach to Animal Liberation* (New York: Peter Lang, 2014); Margaret Robinson, "Veganism and Mi'kmaq Legends," *Canadian Journal of Native Studies* 331, no. 1 (2013): 189–196; Dinesh Wadiwel, *The War Against Animals* (Leiden: Brill, 2015); Alexander Weheliye, *Habeus Viscus: Racializing Assemblages, Biopolitics, and Black Feminist Theories of the Human* (Durham, N.C.: Duke University Press, 2014).

16. Zipporah Weisberg, "The Broken Promises of Monsters: Haraway, Animals and the Humanist Legacy," *Journal for Critical Animal Studies* 7, no. 2 (2009): 22–62.

17. Weheliye, *Habeus Viscus*, 10.

18. Ahuja, *Bioinsecurities*, viii.

19. Belcourt, "Animal Bodies, Colonial Subjects."

20. Kyla Wazana Tompkins, *Racial Indigestion: Eating Bodies in the 19th Century* (New York: New York University Press, 2012), 1.

21. Robert Ji-Song Ku, Martin F. Manalansan IV, and Anita Mannur (eds.), *Eating Asian America: A Food Studies Reader* (New York: New York University Press, 2014); Martin F. Manalansan IV, "Cooking Up the Senses: A Critical Embodied Approach to the Study of Food and Asian American Television Audiences," in Mimi Thi Nguyen and Thuy Linh Nguyen Tu (eds.), *Alien Encounters: Popular Culture in Asian America* (Durham, N.C.: Duke University Press, 2007), 179–193; Anita Mannur, "Food Matters: An Introduction," *Massachusetts Review* 45, no. 3 (2004): 209–215.

22. Manalansan, "Cooking up the Senses," 101.

23. Carol J. Adams, *The Sexual Politics of Meat: A Feminist-Vegetarian Critical Theory* (New York: Continuum, 1990); Carol J. Adams and Josephine Donovan (eds.), *Animals and Women: Feminist Theoretical Explorations* (Durham, N.C.: Duke University Press, 1995); Gary L. Francione and Anna E. Charlton, "Veganism without Animal Rights," *The European*, July 13, 2015.

24. Will Stockton and Karen Tongson, "Triggers and Lions and Vegans, Oh My!: From a Comment War to a Conversation about Cecil and the Ethics of Eating." https://bullybloggers.wordpress.com/2015/08/27/veggies mackdown/, 2015.

25. Stockton and Tongson, "Triggers and Lions and Vegans."

26. Alexis Shotwell, *Against Purity: Living Ethically in Compromised Times* (Minneapolis: University of Minnesota Press, 2016). E-book.

27. John M. Coetzee, *Elizabeth Costello* (New York: Viking, 2003).

28. We are grateful to Anita Mannur, who used this term in her pre-publication review of the *Messy Eating* manuscript.

29. Val Plumwood, "Being Prey," *Utne Reader* (2000): 56–61.

30. Kim, *Dangerous Crossings*, 182.

31. Harlan Weaver, "'Becoming in Kind': Race, Class, Gender, and Nation in Cultures of Dog Rescue and Dog Fighting," *American Quarterly*, 65, no. 3 (2013): 689–709.

32. We are grateful to Chad Lavin for encouraging us to include in the Introduction a more explicit and elaborate discussion of the professional lives of scholars.

CHAPTER 1

Turning Toward and Away

Cary Wolfe

Cultural critic and theorist Cary Wolfe began thinking about human–animal relationships as a student when he encountered animal rights activists on his college campus. This experience resulted in an intense sense of "not being able to turn away," and Wolfe became a vegetarian and devoted activist in his own right. During the same period, he began developing the nonhuman problematic as an object of legitimate scholarly inquiry. On the subject of eating, Wolfe observes that food is a multidimensional and complex problem, shares how his vegetarianism ruined one Thanksgiving dinner, and notes that people have "different kinds of investments" in food. We interviewed Wolfe by Skype on July 14, 2016.

SCOTT CAREY: Could you tell us a bit about yourself? This could include when and where you were born and raised; your formative cultural, intellectual, and political experiences; how you became an academic, whatever you'd like to share.

CARY WOLFE: Well, that's actually three questions! If I forget any of them, just remind me. I was born in South Carolina and lived there until I was about four or five years old, when I moved to North Carolina.

That's really where I grew up; it's where I went to high school and did my university work. Even though I was quite prepared to leave and kind of wanted to go somewhere else, I got this great scholarship to go to the University of North Carolina at Chapel Hill. So I went to college and did my master's degree in Chapel Hill, and then, again, I was looking all over the country for the right PhD program. I had to decide whether I was going to do a PhD or an MFA, because I was both a poet and a scholar and was trying to figure out which one to pursue. I decided to do a PhD, so I was prepared to go anywhere else, again. But just at that moment Duke University experienced a kind of renaissance in the English Department and the Program in Literature; they hired all these fantastic people almost overnight. And I was living ten miles away, and Duke and UNC students could study back and forth with no extra tuition at both universities. So I ended up staying there again and doing my PhD at Duke, which was a fantastic experience.

But let me back up a second. My parents met as high school English teachers in an urban environment outside of Charlotte, North Carolina, in the 1960s. So I grew up in that sort of household—where there was a lot of literature, a lot of '60s culture. Not really of the hippie variety, but more of the informal, liberal '60s environment, even in the South at that time. Then my dad eventually became an academic. He first worked for the North Carolina Department of Public Instruction, where he was the head of their English and Foreign Languages division. He got a PhD, strangely enough, at Duke and went on to be an academic in English education at Old Dominion University in Virginia, specializing in the relationship between writing and learning, especially for kids from kindergarten through twelfth grade. So I grew up in an academic household; I was always around literature, poetry, and music, and art to a lesser extent. To jump back to my experience in school, my graduate school training was really a combination of Marxism and pragmatism. I did a lot of work with Fredric Jameson, Frank Lentricchia, Barbara Herrnstein Smith, Franco Moretti, Terry Eagleton, and a long list of other great people.

But one day when I was in graduate school, I was walking through the Pit, one of the central areas of the UNC campus, and I stopped by the student animal rights group table. And they had all these pamphlets and things displayed, including some material about biomedical research that was taking place on, I believe it was, primates, on the UNC campus. I just idly stopped by and started looking at these materials and I remember saying to the people at the table—who later became very good friends of mine—"You guys have got to be kidding. You're making this stuff up, right?" And they said, "No, it's going on in that building right there." And I was in disbelief. I realized much later, especially with my own work with students, that I had been completely in the dark about this entire infrastructure of exploitation and violence toward animals, and it never occurred to me that it existed, really. I wouldn't call it a "conversion" experience necessarily, but it was definitely an intense ethical sense of not being able to turn away from what I had learned.

So I got really involved in the animal rights movement. And in that part of North Carolina, the so-called Research Triangle, there are a bunch of colleges and universities, including Duke, UNC, and NC State, so there's a high concentration of PhDs—in fact the highest concentration of PhDs per capita in the United States. So there was a very strong state animal rights group, and that was the core of it. I became very involved in this group called the North Carolina Network for Animals; I was in the newspaper and on television and at protests. I was a hardcore activist for a number of years.

But the interesting thing is that while that was going on, I realized more and more that my academic training in Marxism and pragmatism really didn't have much to say about the plight of nonhuman life. That eventually became a very productive divergence; it's what led me to help invent, over many years, what would later be called animal studies, or human–animal studies. Because at that time there was no freestanding theoretical vocabulary that was taken seriously with regard to nonhuman life. In the academy, if you were talking about animals it was taken for granted that you were talking about some sort of symbol or metaphor for a human problematic. In many cases this was in the form of the grotesque, or monstrosity—like Dracula, that type of stuff. But it wasn't a freestanding theoretical problematic in the way that feminism clearly was, or that queer theory became. It took a number of years to build that vocabulary and, in a kind of "Trojan horse" way, bring it into the ivory tower. Fast-forward twenty-five years and it's a different world. People in all different kinds of disciplines take seriously how we relate to nonhuman life—how we represent it, how we talk about it, and so on. So that's the relationship between my official academic training and my background and how I got here.

And I remained active in that way for a long time. When I took my first professorial job, at Indiana University, I was on the Board of Directors for the county Humane Society there. There wasn't really a formal animal rights group in that area, so that was the center of the humane community. Through that I ended up being the Humane Society's appointee to the city's Animal Control Commission, which was a very interesting experience, to say the least. I continued doing that kind of work until I moved to my second academic position, at the State University of New York, and at that point my academic life was becoming increasingly time consuming and demanding. And honestly, I was burned out. This is what happens to activists. The people you're opposing, they get paid to do what they do as an eight-hour-a-day job. What you're doing, you're doing after hours on top of your regular job. So I reached a point when, between the increasing demands of my academic life and being burned out by my activist life, I just decided, I'm putting a whole lot into this now in a different way, in my writing, lectures, and teaching.

In terms of growing up as a southerner in the United States, the thing I want people to understand is that the South is actually a really heterogeneous region. The part I grew up in—the so-called New South or Upper South, the North Carolina–Virginia bandwidth—was actually pretty progressive politically when I was a kid. North Carolina was a solid, progressive Democratic state for decades and decades, going all the way back to a lot of legislation that came out of World War II that helped the textile mills and tobacco farms. I think they'd had only one Republican governor in forty years when I was a kid, something like that. And other parts of the South are the same way. The part of the South I grew up in, especially because of the universities and everything, was a lot more like the rest of the country in many ways.

I had a roommate in college in North Carolina who was from Camilla, Georgia, which was about as far south as you can go and still be in Georgia. He got married when I was in college; I went down there for his wedding, and I had never been to the Deep South. I was actually in shock, because it was so unlike the part of the South that I'd grown up in. I went down there and I said, "Oh, I get it: five rich white families run the county and everyone else lives in shacks." It was almost like going into a feudal environment. The takeaway from this is that, for a lot of people who haven't lived there, it's hard to realize just how different various parts of the South are from each other. So I didn't have a stereotypical, quintessential southern upbringing in that sense. Although my grandfather and great-grandfather—my grandfather, 105 years old, and still alive—they were both Baptist preachers. So I had this interesting mix of some really traditional elements and some really progressive elements in my background.

SAMANTHA KING: Did your animal activism mark your first foray into activism, or did you do other kinds of political work before that?

CW: No, that was my first experience as an activist. It really was an ethical leap. I didn't really have a fully formed and articulated platform that I was working from. It was more of a "hold your nose and jump off the cliff" situation, in which I didn't have a fully articulated ethical platform, but I knew that this stuff was wrong, and it bothered me enough that I wanted to do something about it. All the complex theoretical apparatuses I developed came later, beginning in the early '90s and continuing to the present. That was my first and really only experience as an activist, but it's not one that you forget. I have a soft spot in my heart for activists because I understand what they do and what it takes. Even if I disagree with them, I respect them because I understand what's involved.

SC: What do you see as the primary purpose of your academic work—why do you do what you do?

CW: That's a really hard question. I do what I do for lots of different reasons. One reason is that I really enjoy it—that is, my research, scholarship, and writing—and find it very fulfilling. I enjoy working with students and graduate students and watching the changes they go through. And not just in relation to the ethically complex or hot issues we're talking about today. If I had to sum up what I try to do in my work, including work with students, I'd see it as a process of existential exposure: You have to be very careful not to over-manage or over–stage-manage. Especially through my work with students, I have realized that people come to these issues, and work through these intellectual and ethical questions, very much on their own timetables. Over the years, I've had some students who were extremely resistant to questions involving animals and ethics; often they were raised in a very traditional situation with very traditional viewpoints. But I'll see them five, six, seven years later and they'll tell me it took them years to work through for themselves the relationship between what they do in their everyday lives—what they eat or what products they use—and the kinds of ethical issues we are talking about here. For that reason, teaching really is an art form, especially with these kinds of hot issues like animal rights, or reproductive rights. As a young professor I had to learn a lot about how you can push students, but also how much space you need to give them in terms of their own relationship to these questions. Because a lot of students will just shut down if you put it all out there at one time, very intensely; it is almost too much to process. There is a kind of paralysis that students can experience when they start trying to confront how everyday life—the shampoo they buy, cleaning products they use, food they eat, or car they drive—is completely wrapped up in this structure of violence and exploitation toward nonhuman life. It's almost too much to deal with and then they just say, "To hell with it. I can't think about it. I'm not going to worry about it" and they just go back to the status quo. What I try to do in my teaching, and in my research, is create an experimental space, a laboratory space, in which this kind of existential and intellectual transformation can happen, but is not dictated. The skills you have to use to make that happen, and the settings, may vary widely, but I think that is what characterizes my work with students, and my work as a scholar. I am very much drawn to work that is not about closure, and not about achieving right answers, and not about drawing direct lines between philosophical foundations and political actions that derive from them, but actually work that is quite the contrary. We just did a project in the *Posthumanities* series called *Manifestly Haraway*—Donna Haraway's "Cyborg" and "Companion Species" manifestos, together with about a four-hour-long conversation between the two of us—and I was reminded during those exchanges that this is something Donna and I have in common: how we think about the relationship between the ethical, political, and intellectual dimensions in the work we do.

SC: Thank you. Not everyone speaks to teaching when responding to that question.

CW: Well, I've learned so much about all the things we are talking about from twenty-five years of working with students; I can't imagine discussing any of this stuff without talking about my work with students. There is a lot that I have learned, processed, and then brought into my own theoretical work on the basis of working with students—including my critique of animal rights philosophy in the book *Animal Rites*, and later into my work on posthumanism. That partly derived from my work with students; I recognized that there was something they were resisting in animal rights philosophy that they were right to resist. But they didn't really have the theoretical vocabulary to describe it; they just felt like, "I'm being pushed into a corner here in a way that I don't like." So I took that up and then later worked through it in my own research and in my writing.

SK: Your students' resistance to traditional animal rights theory—was this resistance around its normativity, or its moral prescriptiveness?

CW: I think that was a big part of it. I think another part of it has to do with the specific transferential psychodynamic of the classroom space. Students don't really like being told what to think, and to that I say, "Good for them." So part of the resistance just comes from that. Because if you think about the humanist articulation of animal rights philosophy—in either Singer's utilitarianism, or Regan's Neo-Kantian version—it is a kind of "if P then Q" propositional structure. And at a certain point you can see the students saying to themselves, "That's not how ethical life is. It doesn't work that way; people don't live that way." This is where, if you compare what counts as ethics from the humanist side of animal rights philosophy—let's say from the Singer/Regan line—with the absolutely opposed account of ethics from contemporary continental philosophy and poststructuralism, Derrida for example, people often think, "Well, Derrida is the crazy outlandish contemporary French philosopher who is all esoteric." But actually Derrida's account of what ethics is and how it works is much more in tune with what the students think ethical life is, and that is, as Derrida puts it, "confronting the ordeal of undecidability" in every ethical instance without having a formula in your back pocket that you can just whip out, prescribing what is the right thing to do in any particular situation. So it is actually the contemporary continental philosophers—and not just Derrida, but I would say Foucault, and Lyotard, and others—whose descriptions of ethics actually match up with what the students think about the complexity of ethical life. I think that is a big part of what they were resisting: the reductive, oversimplifying, and propositional analytical nature of the ethical argument from the humanist side. Something in them was saying, "No, this actually isn't how people confront, think about, and live through ethical challenges and complexities."

SC: Thank you. Speaking of complexities—you've been a leading figure in the development of posthumanism as a philosophical and ethical framework. Could you discuss how you continue to find this approach useful, and what you see as the promises and/or constraints of posthumanism?

CW: I'll answer this partly in terms of my own work, but also in terms of being editor of the *Posthumanities* series at the University of Minnesota Press. First of all, we can all agree that posthumanism is an official academic logo now. It's like Coke or Pepsi, or Miller or Budweiser. Forces of academic incorporation being what they are, that's true of posthumanism and that's true of animal studies. There is nothing I can do about it; it doesn't matter what I think about it—that's just the way it is. Something we could discuss at length from a sociological point of view is that academic knowledge in higher education depends very directly, as Niklas Luhmann and other people have observed, on the production of novelty. If we all agreed, "You know what, we got the posthumanism, that's the final word on everything, we're done," then none of us would have any reason for being, none of us would have jobs. So posthumanism as a concept, as a logo, is caught up in the broader economy of academic knowledge production.

Having said that, what I like about the term *posthumanism* is precisely that it is capacious enough, and it is not particularly prescriptive. This goes back to my earlier answer about how I work: I think it creates a space in which you're not prescribing how people must think about any particular question. You are drawing attention to the fact that whatever it is we're doing as scholars, you always have to think about it not just on one level, but two levels. The first level is what is the ostensible content of the problematic that you're working on. In that regard, animal rights is obviously an anti-anthropocentric position. But on the second level—how do you *think* that anti-anthropocentrist animal rights philosophy ends up being quite traditionally humanist in some ways. What I like about the framework of posthumanism is that it creates a space in which whatever we're doing, whatever questions we're taking up, they have to be confronted on both of those levels. Not just content, but also the theoretical, methodological level of how we are thinking these questions.

The other thing I would say with regard to how the "post" in posthumanism is functioning is that it's not a rejection of humanism, which I think would be a really facile position to take. There are many aspects of the legacy of humanism we may find admirable—for example, some ethical imperatives we have inherited from it. What posthumanism does is draw attention to the fact that how those imperatives are taken up on the theoretical/methodological level by humanism actually undercuts and short-circuits these imperatives, and animal rights philosophy is a classic example. You want to say "Yeah, we all agree we should figure out how to value nonhuman life." But the way animal rights philosophy does that within a humanist framework—whether you're talking about Singer or Regan—actually ends up reinstating a very familiar normative picture of what subjectivity is, and it's only on that basis that nonhumans have moral standing.

So, to summarize some of the positive aspects of the posthumanism "rubric," there is a way in which it's intentionally contentless! It simply allows a space in which these two imperatives always have to be taken up—whether you're talking about work in feminism, animal studies, queer theory, environmental ethics, or anything else. This is why for me, as I say in *What Is Posthumanism?*, animal studies is just a subset or a subproblem of this broader challenge of what it means to do posthumanist work. I like that capaciousness and lack of prescriptiveness about it. If you look at the list of titles in the *Posthumanities* series, I think this is what you find. You see work from people who are self-identified object-oriented ontologists; you see work from people like Donna Haraway and Vinciane Despret, who are coming at the relationship between human and nonhuman life from a very important commitment to feminism and gender. You see Mick Smith's book *Against Ecological Sovereignty*. So, as a series editor, what I wanted was a context and a rubric that were capacious enough that the crosstalk between books which sometimes appear completely unrelated can continue to grow, and build, and evolve, and emerge, as we publish more titles in the series. We've just published the thirty-ninth volume, and I think that's exactly what we've seen happening. All the books look different if you go back and read them a few years later, in light of other books that have come out in the series since. That's a way, I hope, that the "logoization" and branding of posthumanism can continually be destabilized, and undercut, and the framework can be kept fresh.

SK: Yes, I certainly appreciate that aspect of the series. Shifting gears a bit to your personal practices—can you describe your approach to eating or otherwise consuming animal products?

CW: Sure. When I was an animal rights activist—I think it was in 1988—I became a lacto-ovo vegetarian, and I have been since. I have never been a vegan and actually, I've never been tempted or felt compelled to eat a vegan diet. Five or six years ago, I started eating some fish again, mainly because I travel so much for work, and I'm taken out to dinner so much by people in all the different places I go. I just reached a point where I realized it was easier on them and on me if we weren't always in the position where I've got to go somewhere where I can have a salad or eat the dreaded "pasta primavera." Which is really just overcooked pasta and overcooked vegetables tossed with olive oil. When

I eat fish now, it's usually shellfish if at all possible. With lacto-ovo products, I try to ensure they come from production processes and environments where you can have at least reasonable confidence that the animals are living humane lives—what I would call the normal lives they ought to be living, which is obviously not the case for industrial food production.

All the stuff I am describing is a lot easier to do now than it was twenty-five or thirty years ago. When I first became a vegetarian, if you didn't live in a city, or in a college town that had your stereotypical "health foods" market, you were in trouble. It was not that easy to do. Now, of course, you can go anywhere—I mean almost anywhere at least—in the States and I think in Canada too, and there are entire sections of the grocery store catering to people who don't eat meat and have other kinds of dietary preferences.

For me, this is in a context of realizing that food and diet are a really multidimensional and complex problem: psychically, existentially, affectively, politically, ethically. In terms of your impact on nonhuman life, what you eat leaves everything else in the dust in terms of impact; that is demonstrably true. But that actually turns out to be just a small part of the story when it comes to what you eat, and why. I know people who don't eat meat, not because they care about animals, but for environmental reasons. I know other people who don't eat meat simply because aesthetically and viscerally they don't like eating meat; they find it gross. There are so many different dimensions to how and why you consume what you do, so I always felt that in my relationship with food, my own ethical principles are always unfolding in a context, and are always only part of the issue. And people who travel a lot, especially in non-Western countries, confront this all the time. They might say, "I'm going to some part of the world and I haven't eaten meat in twenty-five years, but if I go there and I'm served this for dinner and I don't eat it, I'm seriously offending my hosts." So it's just a way of reminding ourselves that drawing lines, as some "beautiful soul" vegans sometimes do, takes place in a context. As Derrida pointed out long ago in what is still a really important and fantastic text, "Eating Well," in the industrialized West, "normal everyday life" is absolutely *predicated* upon an inescapable structure of violence against nonhuman life, one that we are all implicated in, vegans included. That doesn't mean that somebody who eats at McDonald's three times a day isn't doing more violence than somebody who is a vegan—they are. It simply means that it's not differences in kind that we are talking about here; it's a continuum that takes place within this larger context. On the other hand, I think David Wood's wonderful critique of Derrida and "Eating Well" in his essay "Comment ne Pas Manger" is relevant here. He basically says that Derrida may be too quick to generalize all forms of violence as being somehow equivalent. So Derrida can write all this stuff about nonhuman creatures and then say, "Let's go eat steak tartare for dinner." David's point was that there are qualitatively different kinds of violence that take place in these different practices, and they can be identified with some precision. So it's not about being clean versus not being clean; it's about confronting the complexities of these gradations in an ongoing way, in your everyday life—and always in a situation that is very thickly contextualized. That's how I think about the relationship between my principles, choices about what I eat, and the larger context in which those are carried out.

SK: I sense some ambivalence in your last answer, in terms of consuming animal products—could you speak to that?

CW: I would be happy to. My feeling is, in the culture I live in, the world I live in, I can eat like a king without killing animals. That's not the case for other people, in other situations. They live in a different context, and their relationship to nonhuman life, and the taking of nonhuman life, for the purposes of consumption and survival, is very different from my own. It's one of the great benefits of living in the industrialized West: I have the ability to make myself feel a lot better about the violence that I do to nonhuman life because I have access to choices that other people don't have. This is something Donna Haraway and I have talked about a lot and have a point of agreement on, and it also jives with my experiences with students over the years. The issue in many ways is not, "Will death eventually befall the various creatures that we are talking about?" because the answer is yes. If I had to sum up in one sentence the point of view I have gleaned from twenty-five years of teaching undergraduate students: For them, it seems that whether an animal dies is less important than the animal's quality of life when they *are* alive. Similarly, for Donna, the real issue is that you shouldn't institute a discursive technology or any other kind of technology that automatically, taxonomically makes certain forms of life killable but not murderable, simply by virtue of that designation. As she puts it, the issue is less "Thou shalt not kill"—because killing is, after all, unavoidable—than "Thou shalt not make killable" by some kind of taxonomic or—to put a finer point on it, coming out of biopolitical thought—racial designation. In my own relationship to these kinds of questions, this is what I have tended to focus on: that I live in a situation and culture where I have the luxury of not having to kill animals or have animals killed for food. I'm convinced that it's possible to eat eggs and dairy products from animals that are not being made killable for that purpose—and who are in fact sometimes, at considerable expense, being afforded the kinds of lives they ought to be able to lead, in terms of exercising their own capacities, potentialities, and desires. I think that it's possible to do that.

When it comes to the question of eating fish, I feel a lot less bad about eating shellfish than I do about eating a piece of salmon, or some other kind of fish. Why? Because I'm convinced that by-and-large, shellfish don't know what's going on, and a salmon, or a trout, does know what's going on. And there is a complicated section of my book *Before the Law* about this. By "going on," I mean having the capacity to care in some kind of self-reflexive way about what's happening to you and being done to you, something that is not just a stimulus-response sort of mechanism. So for me qualitatively, ethically, there is a difference between eating a salmon and eating a clam. That doesn't mean we can go and kill all the clams. That doesn't mean that clams are now made killable, that we should treat them as simply a brute "resource." It means I try consistently to think about the different forms of life, what they require, what they care about, and how my actions affect them. Of course, we know from the history of science, as I point out in *Before the Law*, that those designations are constantly going to change and shift. In my own lifetime they have moved dramatically in terms of how we think about nonhuman life. So these are some of the things that I think about in terms of this ambivalence—and we haven't even talked about the environment, or labor issues, or racial issues, that are bound up in industrial food production, which we could get into for hours.

SK: That's interesting. I've been thinking a lot about seafood and fish for the last couple of weeks because I'm going on vacation next week to the Maritime provinces, and there will not be vegetarian food available in many of the places we are visiting. So when I look at menus, I'll be thinking of your salmon/clam example.

CW: It's going to be like "fish with fish," I would guess. To come back to this question of taxonomy—that's where a term like *seafood* is just as blunt an instrument as the distinction between human and animal. Which is, conceptually, such a blunt instrument as to be of no use whatsoever. What we are really talking about here is different forms of life, and different ways of being in the world that cannot be described adequately by these kinds of terms. Whether we're talking about seafood or other forms of nonhuman life—and this is where I completely agree with Derrida's approach to these questions—we're really talking about thickening and complexifying the terrain in which we think about nonhuman life. And in that terrain I have a lot more in common with an orangutan than an orangutan has with a starfish, or with a mosquito, or even with a rodent. So clearly the vocabulary *human* and *animal* is of no use in describing this complex topography of different ways of being in the world, and how those affect the kinds of ethical choices we're talking about.

SC: It's certainly complex terrain. Could you tell us a story about a time where your dietary practices have been the subject of awkwardness, celebration, or hostility?

CW: Yes, actually. My number one example is when I became a vegetarian and went back home for Thanksgiving dinner with my parents for the first time, and my mom was so upset. I mean she was almost in tears. I couldn't figure it out, because I had been thinking through all these animal rights arguments. For me, none of it was personal at all. It was philosophical in the most abstract sense: "Here's why I no longer eat animals, and I can walk you through it all." But for my mom it was deeply personal, and she was offended. I gradually realized that to her, it felt like an attack on her being a good mother. This goes back to what I was talking about earlier with the different kinds of investments people have in food, and in sharing meals, and especially traditional meals. It was an important experience for me, because a kind of light bulb went on; I realized that whenever you are talking about these ethical issues around food and eating, the worst kind of fantasy you can have is that they all make sense. I saw in my mom's response things that, to me, were just completely irrational. Yet to her, this represented a violation of a mother–son bond symbolized by, and built around, sharing certain kinds of food that she, since I was born, had lovingly prepared for the family on these special occasions. That was a really important learning experience for me. She eventually got over it, and of course now I've gone back many times since, and in fact last Christmas I made a full vegetarian holiday feast, and everybody was happy and it's like "hahaha." But the first time we went through this, it was really traumatic.

I've also noticed something else, at the other end of the spectrum. When I travel and give talks and do seminars, people are always taking me out to dinner, and often they are really uptight about it; they will ask, "Would you be offended if I ate this?" I get this kind of response all the time, and there is a way in which it is understandable for people who have read my work; they are being polite. But it goes back to what I was saying about students: I just tell people, "Look, relax. I'm not going to tell you what to eat. These are things you have to come around to in your own time." If you want 8 million reasons, you can go read my stuff, but 8 million reasons is never going to equal your doing the right thing either. I confront this all the time—this kind of trepidation—often about taking me out to dinner, and whether I'll be offended if they sit across the table from me and eat a pork chop. And my feeling is, "Just shut up and order dinner and eat, okay?" Now if they told me I could not have a glass of wine, then I would be upset!

SC: I'm sure many of us share that sentiment! We've discussed the relationship between your work and eating practices; could you tell us if—and how—your work has shaped your relationship with companion animals, or vice versa?

CW: It probably is belaboring the obvious, but it has so deepened my relationship to the nonhuman creatures that I share my life with. Partly emotionally, partly affectively, although in a sense, that's the easy part. I think the emotional, affective bond that most people naturally have as children with nonhuman creatures in a way is always already there.

I think what it has done for me more is to make me take the challenge of understanding nonhuman life more seriously. When I lived in Indiana, I became very seriously involved in dog training (this is part of how Donna Haraway and I bonded) and I worked with one of the top dog trainers in the country. He was one of the U.S. Navy's top canine people, but he also worked with the Navy's marine mammal project in San Diego. So he worked with dolphins, sea lions, and some other kinds of creatures doing all sorts of interesting work like underwater bomb detection and dismantling. And he worked with dogs doing search-and-rescue stuff, bomb-sniffing—all kinds of stuff. So I got really involved working with him during that period when I taught at Indiana. Through this work—and this is one thing I think both Donna and I shared—I got schooled in animal behavior, and learning how to perceive the world, or trying to perceive the world, as a particular form of nonhuman life experiences it. The experimental nature of the training scenario will confront you time and again with your own anthropocentric biases about how the world is experienced. If you don't acknowledge that, you won't be a very good trainer and you just won't get very much done. You're forced to deepen and really decenter your understanding of how different creatures, including human beings, experience the world. And that "schooling" happens through both working with animals and also reading about animal behavior, and studying evolutionary, and ecological, and biological materials about nonhuman life.

So all of that has done more than deepen and complicate my understanding and appreciation of how creatures like dogs experience the world, and why they do the things they do, and the forms of affection and understanding I can share with them that I couldn't share otherwise. But more important, as I say in *What Is Posthumanism?*, the final quarry in some ways is that being in relationships with animals also changes how you think about what being human is, and the things you take for granted about what being human is. One way that *What Is Posthumanism?* opens away from the animal question is in bringing in disability studies; that is what the whole chapter on Temple Grandin is about, that we should probably be reluctant about generalizing how *human beings* experience the world. I think this comes back to the question you asked me about posthumanism. You might start out with something called animal studies; you might start out with thinking about the entire evolutionary and phenomenological relationship between human life and canine life. But you eventually loop back to a real reluctance to generalize about this thing we *call* "human," as Derrida puts it. That's been a real side benefit of all this, that began with my saying, "Wow, why do dogs do this?" But it eventually circles back to this question of the human, and trying to think about other ways to talk about how creatures like us get on in the world.

SC: You said you have two dogs right now, but have you always had dogs?

CW: I've always had dogs and cats pretty much my whole life, although it was really only when I got involved in animal rights activism that I would say my relationship to my companion animals changed dramatically. Up to that point in my life, I had dogs and cats the way that most people do. Most dog owners, and I'm going to say this now as a former serious dog trainer, don't actually understand anything about canine behavior. They think that the dog experiences the world more or less the same way that human beings experience the world. This is one reason so many adult dogs end up in shelters. And I was that person until I got involved in animal rights activism, and later in working with the Humane Society in Indiana. My relationship to my companion animals really changed and deepened and became something that had a lot more gravity and texture to it precisely because I took it a lot more seriously. I thought a lot more about it. I just learned how to recognize what was going on around me with these nonhuman creatures that I shared my life with.

If I had my way, I would probably live out in the country and I'd have fifteen dogs and thirty cats. I would also have birds, if I could find the right kind of birds that would be okay with living with me. Most bird species are not great to have as companion animals, but I would love to do that and surround myself with that density and variety of nonhuman life. But I just don't have the life that would allow me to do that right now, too many demands on my time; I travel too much. Even having two dogs is difficult. I don't like to go away from the dogs for more than three weeks, and then I feel like I need to get back and be in their lives again. Which is hard, but that is my cutoff, three, maybe four weeks. I would love to live that way if I could; maybe when I retire I will. Because—if you have read *The Hidden Life of Dogs*, Elizabeth Marshall Thomas's wonderful book—one of the interesting things about dogs, and cats too, is when they achieve a certain numerical density in the group, their behavior completely changes. So one thing that Thomas saw in her book, one dog, fine. Two dogs, fine. Three dogs, fine. The interactions are still within the identifiable companion animal model of a more or less oedipalized relationship between your companion animal and you. But when you hit five or six dogs, she discovered, they're actually no longer interested in you; you are no longer the focus. They actually shift into a pack dynamic, and that becomes the center of their emotional and phenomenological lives, and you are just kind of *there*. The same thing happens with cats, in a different way, when you get into higher numbers. I would love to live in the midst of that and just observe that, but I don't know if I will ever be able to. I hope so.

SC: That would be fascinating to observe, absolutely. One final question, to wrap up: Is there a key dilemma or question that haunts you?

CW: Now this could turn into an hour-and-a-half answer in itself! I don't think I have a *single* dilemma or question that haunts me, in a sort of Derridean, deconstructive way; I tend to think of all of questions of the sort you're describing as being haunted. So then haunting actually doesn't become a unique, or special, or unusual state of affairs; it's where you live always. So for me I would say, you are always in this spectralizing and haunted context, when you think about ethical questions, or political questions, that you actually can't completely confront frontally. By "confront frontally" I mean confront in a way that you could say, "If I were just diligent enough, or just smart enough, or had a good enough heart, that question could be laid to rest and I could be done." For me this condition of what you might call "hauntology," or "spectrality," is a much more generalized, omnipresent, and unsolvable condition. So then you shift from being haunted by a single problem, a particular question, to realizing that if that's the context in which we operate, what we are doing is kind of muddling through as best we can. Taking on what we can, and sometimes turning away from what we can't take on right now. That's the process. So to me there is a permanent, ongoing spectrality and hauntology that's not solvable, and not even fully, frontally cognizable. But we do the best we can.

CHAPTER 2

Subjectivities and Intersections

Lauren Corman

In this interview, sociologist Lauren Corman reflects on her profound interest in exclusions and the potential for both activism and scholarship to acknowledge new and more complex subjectivities. On the subject of teaching in critical animal studies (CAS), Corman addresses the need to develop novel approaches for novel content and her desire to teach her students critical reflection and thinking without provoking defensiveness. She also shares her insights on the growing discipline of CAS and stresses the significance of intersectionality to both its present and future iterations. We interviewed Lauren Corman by Skype on December 16, 2015.

SAMANTHA KING: Since we are interested in the mutual constitution of personal and academic interests, we begin with biography and wonder if you could talk about when and where you were born and raised; your formative cultural, intellectual, or political experiences; how you became an academic; and how you came into the field in which you now work.

LAUREN CORMAN: That's such a great question; it's really layered and difficult to distill. I guess I'll start by saying I was born in the United States but we moved to Canada when I was three. I grew up in a beautiful area of rural Manitoba but a very strange little town. It was dominated by nuclear research. Most people's fathers, in particular, were nuclear physicists or nuclear researchers of different kinds. My family wasn't, but it meant that my primary, elementary, and high school years were taught from a very scientific point of view. There was a real emphasis on the natural sciences, and the humanities and the social sciences were considered not particularly valuable. As somebody who is interested in the humanities and the social sciences, I railed against that position, and I fought that throughout my childhood. As well, I had an abusive father, and although my mom is very supportive and wonderful, my childhood experiences were my beginnings of thinking about injustice and oppression.

In combination with that, I've had a longstanding interest in animals. My mom says that I was always an animal activist, in the sense that I didn't want to go to the zoo, and I didn't want to go to the circus because my feeling was that the animals were sad. Some of those feelings about animals didn't become academic and activist interests until the end of my undergraduate degree. Yet those seeds were long planted; I always felt a connection with my cats growing up and that kind of thing. I thought that I would end up doing English in my undergrad, after I graduated from high school. I had been writing terrible poetry all the way through high school [laughs]. When I began my undergrad, I thought I would end up being a novelist—I'd always loved English. I did well throughout my undergraduate degree. It was affirming to me, given that that town had been quite dismissive of the arts, to see that I could excel within university and be rewarded for doing humanities work. I was just a number. In some ways that was great because I had the anonymity, and then I received the validation that I could write. Strangely enough—even though I've ended up in sociology—I wound up dropping my undergraduate sociology course. I was quite a dedicated student early on, but I dropped sociology. I didn't really understand the professor, I didn't think I would get a good grade, so I was encouraged by my partner at the time, and my mom, to take gender studies.

That was the beginning of my politicization. I embraced gender studies, and I think I became a force to be reckoned with at the University of Manitoba, where I was doing my undergrad. I started learning about issues of oppression, which both opened my eyes and also gave me a language to describe my own experience, and also the experiences of people around me. I really threw myself into it. This was the mid-1990s. Issues of intersectionality were at the forefront of gender studies, and there was a lot of interest accounting for the kinds of exclusions occurring within the women's movement. That's how I arrived in politics, thinking about how issues of race, gender, sexuality, and ability, and those sorts of dynamics have been excluded from Western feminism.

Adjacently, I also got involved in the punk scene in Winnipeg. Within the scene, there was a dovetailing between the political work I was doing on campus and the theoretical ideas I was learning, and my interest in punk. It was a very progressive scene that was doing deeply intersectional work, although we might not have called it that. It was through the punk scene that I was first introduced to animal issues and exposed to issues such as vivisection at a punk show, by a local band named Propagandhi. They had a bunch of different local groups provide information at their show. There was a group called People Acting for Animal Liberation (PAAL); they were playing a video of a monkey being vivisected. That was a crushing moment, and it was really terrifying too. There was this monkey who was screaming, and I guess that was the moment of connection between the animals I eat and this particular animal who was clearly suffering. I went vegetarian that night, and then pretty quickly vegan after that. I was interested and I started learning about industrial forms of animal use.

Here I was in gender studies being radicalized and coming to a greater political consciousness, but I was really frustrated that animals couldn't be part of the conversation; that for all of the exclusions we were speaking about and all this emphasis on intersectionality, nonetheless, where was speciesism? We didn't have a language for talking about it. I felt like the women's movement and feminism, by its reluctance to take up the question of the animal, was reproducing similar kinds of exclusions that it had been so diligently trying to address in the 1990s. I reluctantly left gender studies to pursue my master's and then my PhD in environmental studies, because at least in those fields I felt there was room to talk about the nonhuman. That's how I ended up going to York University and doing my master's and my PhD in environmental studies. There wasn't a critical animal studies focus at the time; there was a little bit of animal studies—human–animal studies—but it wasn't an area that you'd get a degree in. Actually I thought it would be funny to do a degree in animal studies because people would ask me why I was doing environmental studies, and I would say, "Well what am I going to do, 'animal studies'?" This was laughable at the time.

I felt like I did two concurrent degrees because I was interested in animals, but they didn't really fit within many environmental studies frameworks, which are really informed by a kind of interest in species and endangered animals, and a particular understanding about animals that isn't about individuals. Certainly there wasn't much room for domestic animals. That's starting to change. In fact, domestic animals are kind of the shadow side of wild animals, who are prized within environmental studies. They've become the new "other," or a kind of continuation of that othering that we see in Western societies. That's why I did the radio show, in part, because I was quite desperate to have conversations with other scholars and activists who are doing animal work. I did that show from 2001 to 2009, until I got my job at Brock University, and a bit into 2010. I was specifically hired at Brock in 2009 to teach critical animal studies, and I believe part of the reason why I was hired was that I do intersectional work. I bring a feminist consciousness forward, which is dedicated to intersectionality, and I think that made sense to the sociology department. A lot of people do that work in my department. I also had about 300 interviews under my belt at that time through the radio show. I had a pretty good understanding of the scholarly and activist landscape at the time. It's a big question, but that's my attempt at answering.

SK: Could you tell us about the primary purpose of your academic work, and your motivations for engaging in it?

LC: It's shifting a little bit, but primarily I would say that I'm interested in questions of human and nonhuman animal subjectivity. That's partially because of my interest in social justice movements, and then locating the animal movements within that framework. You can see these interests unfold through my dissertation research, the radio show, and the work that I did on my master's in which I was interviewing slaughterhouse workers and interested in the exclusion of slaughterhouse workers' voices within labor histories and within the animal movements as well. I became interested in, or I became aware that many social justice movements in the West—and I think that the animal movements also mirror this—begin with an interest in moving those who are oppressed from a category of object to a category of subject. That's true for the women's movement and it also rings true for various queer movements. We see it as well, in Canada, in regards to the Indigenous movements. There is a desire to bring into the public sphere richer versions of subjectivity for those who've been oppressed, as a movement against the objectification and dehumanization of these groups. The animal movements are also engaged in that work, and I'm centrally interested in that work.

But it's not just about subjectivity; it's also about subjectivity in relationship to intersectional analyses. My beginning in terms of my political consciousness, and my theoretical work, was an attempt to think about issues through an intersectional lens, and that intersectional feminist perspective still informs all of my work. I want to talk about marginalized groups and, in particular, emphasize their subjectivity, but do it from an intersectional perspective. I'm interested in this trend toward talking about marginalized groups and figuring out how to do that work without reproducing victim discourses, or a kind of reductionism in their subjectivity where they're just reduced to suffering beings. We can see that the emphasis on agency and resistance, and all these other aspects of subjectivity beyond suffering, has been incredibly important for various different Western social justice movements. The animal movements are lagging behind in that regard, but that's beginning to change. So that's part of the intervention of my work—that is, to emphasize those more complex versions of subjectivity that animals and other oppressed groups possess. That's part of the reason why, over my time at Brock, since 2009, I've brought in much more cognitive ethology—the study of animal minds and behavior—than I had anticipated. This was percolating when I was doing the radio show in an effort to figure out who is talking about animals themselves, outside of just victim discourses. I started to interview more people who work at sanctuaries, and more people such as Marc Bekoff. I did an interview with Barbara Smuts right at the end of my time with the radio show. They could talk about animals outside of a purely victim paradigm. I completely understand why the animal movements have emphasized suffering and victimhood; it's a crucial corrective to the ways that animal suffering is often not seen, not discussed, and intentionally obscured. There's a lot of distancing that happens regarding animal practices, and then people's consumption practices. I do understand why. I think it's important and obviously there's tremendous suffering, but I don't want us to stay there.

Some of the work in cognitive ethology I find especially promising for bringing animals more into the picture, attempting to complement an understanding of their suffering with a larger understanding of their lives in regard to their emotional capabilities, their sociality, their capacity at times for culture, these kinds of things. A lot of those conversations are happening within cognitive ethology and research that bridges the natural sciences and the social sciences. In part, that's where some of my work is going. Again, it's carrying forward that central idea of subjectivity, putting it in conversation with intersectionality, and always being accountable to the multiple ways in which animal oppression is entangled with other forms of oppression. They can't be separated. Then, as part of that mix, bringing in cognitive ethology to further open and deepen our understanding of who animals are outside of these very peaked moments of suffering, or these extended periods of suffering in their lives.

As I say in the chapter I co-authored about critical animal studies and pedagogy: This is not just a purely theoretical idea. It has real political and practical benefit, because I've really seen it in my own teaching. When I talk with students about animal suffering in the absence of talking about their emotional lives or their social capability, it doesn't land in the same way. Whereas if I talk about animals in a much fuller way, when I show them graphic imagery, or I discuss some of the violent ways that they're commodified, that combination of discussing animal emotionality with descriptions of suffering is much more potent. It means a lot more to the students; it seems to elicit much more of an empathetic response, and it just seems much more motivating. It's also just not the regular kind of onslaught of relentless suffering that sometimes the animal movements portray. Again, I say all of this as somebody who is part of a wave; I don't feel like I created this work. They're ideas I picked up on, issues I was trying to intervene in myself, and then build on through scholars such as Barbara Smuts and Karen Davis, and others who have been engaged in this work for a long time.

SK: I don't know if you identify as a sociologist with your interdisciplinary background, but could you talk about what it's like to study animals in a sociology department?

LC: Yes, I feel like I slipped in under some kind of wire, actually. It's strange. In some ways, though, people weren't surprised that I ended up in sociology because, at least at Brock, the kind of sociology that's done here is largely engaged with feminist theory, labor theory, and postcolonial and critical race theory. That's all of my work too; it's just in conversation with animal issues. It's actually quite an interdisciplinary program in the sense of people's theoretical backgrounds; we don't strictly house faculty with degrees in sociology. But it did occur to me the other day—I was asking a friend, "Am I a sociologist?" I wasn't really sure: I teach full time in the sociology department, I'm now a tenured professor in sociology, so am I a sociologist? One of the ways I identify as a sociologist is that I've always been compelled by questions of society and social relationships. As my research is more greatly informed by cognitive ethology, I'm really interested in nonhuman animals' social relationships. To me, it seems like a really good fit for sociology in some ways; it's an extension of, and a kind of disruption of, some of the paradigms within sociology. Like anthropology, sociology has tended to assume that some of its main concepts such as society and culture are exclusively human. Not only strictly human, but also defining the human. But then what makes us human? These concepts of society and culture are really ripe for deconstruction. So in some ways I feel well positioned within the academy right now, although I consider myself more of a cultural studies person. Interest in social relationships and inequality helps define my department, which is, I would argue, a critical sociology department. I continue to bring forward my intersectional analysis, and it feels like a good match with the department.

Where I bump up against some heads, I suppose, is in that questioning of what constitutes society. I developed and have been teaching a course since 2009 called Animals and Human Society, where even the name of the course has been the subject of debate. There's an enduring assumption that society is humanity, or the domain of the human, and that only humans have the capacity for social relationships and what we might call society. Scholars such as David Nibert have been working diligently to deconstruct the kind of humanism that shoots through a lot of sociology, and I feel like I'm also engaged in that work. It feels like a good home for me, but at the same time, some of its fundamental ideas are quite steeped in a humanist position that I find problematic and try to challenge.

SK: Could you elaborate on what critical animal studies pedagogy means in your mind, and why you see this as an important undertaking?

LC: Completing my dissertation, I was primarily interested in questions of voice, animal voices, other marginalized voices, the humanism of the voice metaphor, and the centrality of the voice metaphor to social justice movements. These issues animated my work for so many years. But it's been valuable to reassess and think about my direction post-tenure. I earned my PhD in 2012 and then I applied for tenure in 2014, so it was very fast. During my tenure application I had to look back at my career and my activism and note some themes. I needed to ask myself in a direct and comprehensive way, "What makes me unique as an academic?" Or, "Where do I see my career going?" At least one of those threads relates to pedagogy. The intervention of the radio show was a pedagogical enterprise in the sense that I wanted to bring animal issues into the public sphere and tried to curate better conversations about animal issues. I was motivated to bring complex theoretical ideas into a public forum, making them accessible, and handing them over to people. The radio show, and of course being a teaching assistant and those sorts of things, primed me for thinking about and engaging in pedagogy. But, significantly, prior to this, my undergraduate gender studies professors were graduates of OISE [the Ontario Institute for Studies in Education]; they put questions of pedagogy at the forefront when we came to class. That was great and kind of radical, to have a conversation about pedagogy as part of our curriculum, about how we were going to be taught, how we were going to learn, that pedagogy was something that you could step back from, evaluate, and critically engage.

There was a series of little moments that led up to this interest in critical animal studies pedagogy. Part of my developing interest in pedagogy came from noticing, when I began teaching in 2009, that there weren't many models showing how someone would teach this area. It wasn't like I had taken critical animal studies classes in my degrees and could draw on those approaches, or translate them, into this new context. When I designed courses, I sometimes felt I was starting from scratch. John Sorenson teaches in the department and had been teaching critical animal studies courses for a number of years, but in terms of being able to just teach it in this area, I didn't have a whole bank of courses to consult. Moreover, I wanted to teach based on my own research interests, primarily intersectionality and subjectivity. In other words, I was in this amazing position, as scary as it's been sometimes, in which I occasionally lacked models. Because I was uniquely hired to teach the critical animal studies courses, I spent five or six years just teaching critical animal studies. Last year was the first year that I taught contemporary social theory, which wasn't an animal-focused class. The reason I'm giving you a bit of background is that pedagogy was an ongoing interest, but it was also borne out of necessity, to develop effective teaching.

I was also distinctly aware from my own master's and PhD experiences that talking to people about animal issues is quite emotionally fraught at times and quite volatile, and so I wanted to develop methods for precluding some of the defensiveness that can be there. People will often ask me about my teaching, "Don't you find that students get defensive?" For the most part, not that I'm a magician or anything, people don't get defensive. I've tried to build into my pedagogy a series of ways of precluding that, for example by emphasizing larger systems of oppression and talking about economics, which helps to de-personalize some of the information about animals. People often take animal issues personally because everyone is participating in some ways in these systems, and people can sometimes feel guilty or they can feel upset about it and see it as a reflection of their being a bad person. I always try to put a bit of a wedge in there to encourage people not to see it as a matter of being bad individuals and instead emphasize a larger structural analysis. I'm building on preexisting pedagogical techniques in which people are talking about larger systems of oppression and systemic forms of inequality; I am extending those to show that issues of speciesism and anthropocentrism take similar, and also unique, forms. We can talk about issues of domination and oppression without saying people are evil or bad. It's about larger systems we can name and critically interrogate. This is how I came to some of my pedagogical interventions and motivations.

SCOTT CAREY: In your forthcoming essay "The Ventriloquist's Burden," you offer a critique of exclusions performed by certain versions of posthumanism. Could you review that critique for us and describe how you locate your work in relation to posthumanism? Could you also give us some idea for what you see as the value of a posthumanist perspective for animal movements?

LC: All of these categories or these terms like *animal rights* or *posthumanism* are useful placeholders, but I think they're also quite malleable and fluid. Part of my contention with Donna Haraway's work, such as within her book *When Species Meet*, is that there's a kind of strange rigidity to her notion of animal rights. I don't want to present posthumanism as if it's separate from animal rights and animal liberation philosophy, although I know some people do that. I think Donna Haraway ends up doing that as well, in order to define her own research. Considering I've done animal rights and liberation work for so long, I think that people do all sorts of work under these categories and use them in a diversity of ways. I just wanted to have that noted as an overall statement before I talk about the interventions or how I see myself contributing to posthumanism, because I've seen these as really messy categories. It makes answering the questions a little bit difficult.

I think in some ways there's more overlap than is recognized between animal rights and posthumanism, which are often pitted against one another. But one of the things that I find promising about posthumanism is that, for myself, coming out of gender studies and intersectional theory—and, in particular, poststructuralist and postmodern analyses—there's been a great deal of useful emphasis on deconstructing what we think of as the liberal Western subject, all the different ways that that subject is constituted by a series of unmarked categories. Categories of whiteness or heterosexuality or able-bodiedness—all these different categories are tied up with how, in the West, people think about who is a subject. Many people who have been engaged in intersectional work and poststructuralist and postmodernist analyses have been doing such important work to note the unmarked ways that the subject functions to talk about race, class, gender, and sexuality, etc. To me, that has been very useful work. It informs intersectional theory and helps us attend, in better ways, to how various different groups have been marginalized and how that marginalization constructs the subject. It's not just that different groups are excluded; it's that those exclusions help mark and define who counts as a subject and who is not a subject.

Posthumanism, I think, arrives on the scene in that it helps contribute further to those discourses by asking us to also consider how the notion of the subject is marked by humanism. Humanism is also there within all of these other analyses, but people aren't necessarily recognizing that species is another category that helps constitute who counts as a subject in society. I didn't see the deconstructive work around the subject happening as much within animal liberation and animal rights ethics and philosophy. To me it seems absolutely crucial, and people such as Cary Wolfe show us in quite stark terms that some of the paradigms that people are writing under animal rights and animal liberation are actually predicated on humanist notions of the subject. We will always be mired in a kind of domination of the human if we're not willing to unpack the assumption that what it means to be a subject is to be human. When you have various different groups within the animal rights movement trying to get animals into the category of rights, they're often doing it based on a humanist understanding of who gets to have rights. It's based on capacities such as language use, or rationality, or those sorts of criteria that Taimie Bryant calls the similarity argument or the similarity approach. While that is very useful work to show similarities between humans and other animals as a way to grant animals certain rights, the notion of rights is nonetheless based on a humanist exclusion of nonhuman animals. So we're kind of in this paradoxical moment where the very category of human rights is predicated on not being an animal. So how do you get animals within that rights paradigm? Well, you try to say, "Look how animals are like existing rights holders—i.e., human beings." To me, when I read that critique by Wolfe, that made so much sense and raised concerns about the work that the animal movements do.

SC: Could you discuss some of the problems with humans' speaking on behalf of animals, and how you've tried to resolve that issue in your work?

LC: Looking back on my development as a scholar and as an activist, I see that coming out of the women's movement in the 1990s; there was so much concern with questions of cultural appropriation and voice appropriation. It was understood that the white, Western women's movement had engaged in a series of profound and damaging mistakes in terms of speaking on behalf of others: women in the Global South, women of color, etc. To think about and take seriously questions of appropriation marked the beginning of my development as a political actor. These were really questions and matters of colonialism and imperialism that continue to be significant. There was an ongoing set of debates, which persist, about the issues of power involved with speaking on behalf of someone else, and what it means to be an ally, or what it means to speak with somebody. I think in some ways, those questions were swirling around for me at the beginning of my emergence into political consciousness and theory. However, I was still largely engaged in the kinds of activities that the animal movements have historically done in the West, which involves a pretty intense feeling that we know what is wrong with what is happening to animals and see ourselves as speaking on their behalf—that is, positioning animals as voiceless, and animal activists as their voice.

In my dissertation research, which centrally revolves around this question, you can see that this kind of assumption is built into the animal movements historically and contemporaneously by rendering animals as unable to speak for themselves. I think there's a whole bunch of reasons for that. In part, politically, people are largely correct that animals don't have much of a political voice. That's vital to correct by speaking on their behalf, because we recognize the relative powerlessness that animals have to represent their interests in terms of a public or political intervention. It's not that I think people are being ridiculous for talking about themselves as the voice of the voiceless, which is the central trope of the animal movements. However, given that in a Western context the notion of voice is very tied to subjectivity, there's also a way in which the animal movements render animals as nonsubjects at the same moment they try to be their voice and bring them into the public sphere. That is a problem. It's something that we should be concerned about because it doesn't necessarily imbue a sense of responsibility to try to figure out better ways, or at least to try to engage with the problematic of possibly bringing their voices into this larger conversation. If we are going to seriously grapple with some of these other debates about appropriation and the kinds of colonialist moves that can happen when you speak on behalf of someone else, then we need to build that kind of critical consciousness into animal movements, which are pretty brazen about being able to, or feeling able to, speak on behalf of others.

I did a radio show called *Animal Voices* for almost a decade, and this is the kind of language that's out there in terms of the work that we do as activists: that we must be their representatives. In some ways we *must* be their representatives. There is a certain accountability to do that, but because we haven't engaged in a more nuanced way to think about animal subjectivity and combine that with the idea that we are their voices, we are left with a kind of reductionism. There's a way in which we continue to position animals as strictly victims, reduce their lives to images of suffering, and don't do our homework around what animals can be outside of those contexts, how those contexts themselves are quite complicated, and how animals do attempt to resist. They do have forms of agency. Their lives outside of these contexts can be quite complex and interesting, and they seek out pleasure and all of these sorts of things. Again, the central trope that embodies a lot of animal movements is that humans must act as the voice of the voiceless. It is an understandable position, but it is also one that is worth thinking about more critically. I argue that the animal movements need to centralize animal subjectivities in such a way that is more layered, more honest, and more humble—not only to be fair to the animals but also because it has political import as well. If you engage with researchers who are attempting to know animals on their own terms—for example, people such as Marc Bekoff—then you see that this is in line with other social justice movements that increasingly refuse to reduce others to objects and suffering victims. I see this unproblematized notion of speaking on behalf of animals and being the voice of the voiceless as a potentially colonialist or appropriative way to think about others.

ISABEL MACQUARRIE: You make a concerted effort to bring the voices of activists into your scholarly work. Can you tell us why you do this and whether you've encountered any resistance from inside the academy?

LC: Part of the work that I've done fits well within some of the founding principles of critical animal studies—or, at least, some of the initial ideas that people used to define the field. One of the issues that were central to the folks defining critical animal studies was an appreciation that advocacy and the academy should be in conversation, bolster each other, and hold theory accountable to activism and on-the-ground work. There's a kind of response to, or wanting to resist theory for theory's sake, or abstractions to the point of being largely meaningless to the material conditions that animals face. There was also the idea within critical animal studies that we should value the work that animal liberationists and animal rights folks had been doing and not engage in the conventional scholarly practice of disregarding those voices. Engagement with activists' work was seen as a kind of political move and theoretical move.

It was a political move in the sense of resisting a dismissal of activist perspectives—and it was theoretical in the sense that we stay open to what on-the-ground activists are saying. That should inform our theoretical work. That made a lot of sense to me coming out of the women's movement and gender studies, in which the theories being developed were always being developed, or at least largely developed, out of the women's movement's direct engagement with fighting patriarchy and other issues of oppression. Again, critical animal studies was built on theoretical paradigms familiar to me through gender studies. Then, when I was doing the radio show I wanted to collect all of those voices together because I thought that doing so would give us a variety of perspectives, or a larger set of perspectives on animal issues. If you're going to talk about something such as the Canadian commercial seal hunt, for example, you'll want to speak to activists about the conditions that they're seeing, their experiences, and some of the observations that they're making. That would be vital to the conversation, but you would also want to include somebody who could talk about cognitive ethology and seals. You'd want to bring in somebody who could offer a labor perspective and have a labor theory background who could talk about working conditions on the ice, or what the conditions are in Newfoundland that precipitates people's working in the industry. You would want to interview Indigenous people who profit from the hunt. It would be complementary to have these voices combine in terms of providing a more holistic understanding of any issue. Like I said, critical animal studies was trying to work against a kind of disembodied, theoretical trajectory that isn't accountable to material conditions; that is something the animal movements and animal activists have been trying to ameliorate or stop for a long time. Activists can help keep people grounded in those material conditions and bring their lived experience and firsthand experience to the table. I also think it's more compelling for students. Practically, in terms of my own teaching and my own writing, some of the stories that activists tell about their lived experience grab people much more. There is a kind of affect and passion that people bring to their activism and work on animal issues. It can help flesh out the theoretical perspectives that I'm talking about in class which can tend to be quite abstract when they refer to the animal in this kind of generalized way, without talking about particularities. Something that activists can do is talk about the individuals they've met if they're talking about liberation, and this can be a helpful corrective against the abstraction of the category of the animal that some people engage in within the academy. Those are some of the reasons why activists' voices have been key to my own work.

Generally speaking, I haven't encountered resistance from inside the academy related to this because I'm fortunate to work in a department that values activism and tries to stay accountable and in conversation with activism in terms of people's own research foci. Generally speaking, it fits with that critical sociology perspective. And, like I said, my students really seem to like it. I often feature radio interviews as part of my course materials, or I try to bring in activists' voices on YouTube and things like that, because it gives people some stories and faces to help ground some of the theoretical ideas. Also, I hope it gives students the sense that these are ongoing, lived, and contemporary debates that are happening right now. I always say, "These issues are very much alive today!" to try to show them that this is something people are actively participating in through a variety of means. That makes it more exciting and also, hopefully, more relevant so that we're not just talking about theory for theory's sake. I think that kind of view instills the students with a sense of excitement. That's part of the reason why I interviewed Will Potter with my class after we read his book *Green Is the New Red* for a course I taught this year. We read all of these theoretical perspectives on eco-terrorism and then we talked to a journalist and read his book. We got to interview him about his own personal experiences and his own understanding of activism, and he emphasized the voices of activists in the book. It took some of the more abstract theoretical ideas and helped ground them and make them seem more alive, contemporary, and exciting.

IM: What is your current perspective on consuming meat or other animal products? Have your thoughts and feelings changed over the course of your career on these matters? And if so, how?

LC: I know it's continuous with the other things that I'm talking about, but it's also a shift in register, because it's about personal practice.

Of course, veganism can be, and is for me, tied to these larger theoretical questions and these questions about activism and interventions about animals' lives, but it's something that's very personal, which is part of the reason why I often talk about issues of veganism in a peripheral way with my students. It can be woven throughout some of the readings, but I don't come out saying, "Veganism!" right away. Often the students will bring it up to me, and I see that as a teachable moment. I enter into a conversation in that way. It's a highly personal thing and that's part of the reason why some people are defensive about it too, because we are all implicated. Everyone is suddenly understood as making a choice, one way or the other, about whether to consume animals or not. It's hard to position yourself as neutral.

I notice that I'm kind of shifting my register by not talking theoretically, because this is about personal practice, which is, like I said, attached to these larger theoretical ideas. I mentioned how I became vegan, and it's part of a daily practice of eschewing the objectification and commodification of animals. I think it was Scott's question about what are the major ideas or key concepts that I'm engaged with where I talked about the kind of work that many social justice movements do, which is to try to move a category of beings from objects into a category of subjects. That's at least partially articulated through some of the liberationist critiques of animal industries, which really question the rendering of animals as property. A lot of the work that I've done in my teaching has been about legal issues and the law. The central categories of personhood and property that exist in the law get at the fundamental tension between objects and subjects.

Veganism, to me, is about a kind of direct challenge to the property status of animals and the notion that they can be rendered as objects. It's a kind of daily personal boycott, a rich practice of eschewing the understanding of animals as objects or servants from the beginning of their lives to their deaths. Within an industrial context, everything about their lives is geared toward serving human beings. Their bodies are manipulated in various different ways and mutilated in different ways in order to serve the purpose of being objects. To not consume those products is a way of forwarding a practice and a political position, which is not to reinforce but instead challenge the property status of animals and their objectification. Increasingly, as I've learned more about animals' social relationships, their emotionality, their psychological capacities, and at times even their cultural capacities, I've found my veganism and the practice of my veganism becoming a lot sadder, to be honest. I think that, like many people who are vegan or vegetarian, we're aware of the physical suffering that animals experience within an industrial context; it is so extreme. The violence is just so extreme, and that hits a lot of people.

For me, anyway, delving into the central idea of animal subjectivity, you realize the real complexity involved with the kinds of bonds that are broken, or the kinds of ways that social relationships and their emotionality are harmed through industrial practices. It really is quite devastating. So when people say to me, "Don't you miss cheese?" or those sorts of products, at this point I really think about dairy as a disruption or a kind of mutilation or violence toward not only animals' bodies but also the social relationships between calves and their mothers. There is a kind of anguish that cows and calves experience when they're not allowed to engage with each other. When the mother is not able to mother her baby, that kind of pain certainly transcends any kind of physical—it's not strictly a physical thing. I was already a dedicated vegan; I've been vegan since 1998 or 1997, but it's gotten deeper and probably a bit more depressing as I've learned more about animal minds and behavior. This is somewhat complicated by living in the Niagara area, though, because there's been a lot of significant debates unfolding over the past few years about a First Nations deer hunt that happens near St. Catharines in Short Hills provincial park. It's a First Nations hunt and the animal movements here have been quite divided on the issue about whether or not to protest the hunt. Some people have, and some animal activists have also stood in solidarity with Indigenous people who are trying to support that hunt and enact their treaty rights. It's been important to me—and I don't feel like it is a contradiction—to think about how issues of hunting are different. Hunting for sustenance is different than talking about the kind of industrial commodification and capitalist rendering of animals as property. I don't think those analyses are necessarily transferable to thinking about First Nations relationships with animals, and others who are hunting for sustenance and cultural purposes. But, it is interesting: There is someone I know who is an animal activist and went out in support of the First Nations hunters to stand in solidarity with them. In the end she ended up eating some of the deer meat that was offered to her from that deer hunt. It's not something I would have necessarily done, although I don't know. In that situation, she had a whole series of experiences that led up to her accepting that meat.

I do see meat eating and the consumption of animal products as complicated, and I'm cautious of language that's used within critical animal studies related to this notion of "total liberation," which I really don't like. Even though that phrase is supposed to enact intersectional theory and address all sorts of forms of domination simultaneously, I don't think we always know what that liberation would look like. Sometimes people assume as part of a total liberation approach or in their talk about veganism that these are paradigms that will work for everyone in every context. I feel equally annoyed by people who say that veganism is for everybody, as I do by people who swing the other way and say that "poor people can't be vegan!"; that is the kind of reactivity around veganism that you also see. Well, there are a lot of poor people who practice a plant-based diet, or sometimes eating a plant-based diet can actually be more affordable to people. And, sometimes people are poor and are also animal activists and vegans.

Occasionally, people talk as though they can use these categories to fortify a particular political position. Sometimes it's actually about erasing animal subjectivity, so I don't want to go that way either. But I also want to recognize that sometimes there are very real material conditions that make it difficult for people to access vegan food, or that there are a whole set of cultural legacies that would lead people to want to consume animal products. It is understandable for white middle-class vegans to assume that everybody is going to want to, or should, engage in veganism, when those same people have difficulty fully engaging with issues of colonialism or developing a critical consciousness about colonialism and the importance of food procurement and consumption as cultural practices. But that's not to say that those traditions, or those cultural practices, shouldn't also be exposed to a critical perspective and that there's somehow a truth that gets to exist outside of cultural critique. We would just want to bring that into the conversation about how people are talking about and thinking about the feasibility of veganism. Certainly, Margaret Robinson, who is a Mi'kmaq scholar in Toronto, sees continuity between her Mi'kmaq traditions, cultural understandings of animals, and her vegan practice. But, I'm also cautious to not create a false dichotomy between Indigenous understandings of animal practices and the sets of philosophies that underpin white vegan practices, because they're not mutually exclusive. Certainly there's been more people who are talking about human–animal relationships in ways that are based on mutual understanding and respect, non-object relationships, and non-objectified understandings of animals that come out of Indigenous philosophies and traditions. So, there seems to be a growing realization among non-Indigenous people that Indigenous thought offers a whole series of valuable insights that can help re-think the possibilities of better human–animal relationships. Those are some of my thoughts on veganism lately.

CHAPTER 3

Being in Relation

Kim TallBear

In this interview, Indigenous and feminist science studies scholar Kim TallBear explains that she was raised implicitly to understand that animals have "their own life trajectories" and social practices and that humans live "in relation" to nonhuman creatures. Expanding on the overlaps between Indigenous thinking and recent posthumanist and new materialist scholarship, TallBear suggests that Indigenous thinkers have always engaged a posthumanist understanding of self and other. These thinkers, she contends, are both deeply engaged in understanding the fluidity of relations among the many things that are "living things" and relatively uninterested in the concept of the human. We interviewed Kim TallBear by Skype on September 10, 2015.

SAMANTHA KING: Can you begin by talking about where and how you were raised?

KIM TALLBEAR: I was born in Pipestone, Minnesota, in 1968. My next book is on the Pipestone quarries. It's where a lot of the stone is harvested for ceremonial pipes in the United States, and it's a National Park Service site, so there's a lot of history there. I grew up about fifteen miles from

Pipestone on the South Dakota reservation of the Flandreau Santee Sioux Tribe. A lot of people in Pipestone are enrolled in Flandreau or from another Dakota reservation in Sisseton, South Dakota. So there's a lot of back-and-forth between these three small towns, and a lot of people would call Pipestone a spiritual center because the quarries are there. But there was also a Bureau of Indian Affairs boarding school there that I think closed in the 1950s that makes Pipestone a place with diverse histories. My grandmother went to that Indian school. I was born in the public hospital in Pipestone, so I have a lot of contact with that space, but grew up just over the border in Flandreau and went back and forth between Sisseton and Flandreau where everyone is related. I lived there until I was about fourteen.

My maternal great-grandmother and my grandmother raised me, and then I went back to live with my mother in St. Paul, Minnesota, when I was in high school. But again, there is in general a lot of back-and-forth between the Dakota reservations in the southwestern part of Minnesota and the eastern part of South Dakota and the Twin Cities. The Twin Cities is a really interesting place: It's basically Dakota and Anishinaabe people there, and they go back and forth between the cities and their reservations in the northern part of the state and in Canada. So two very vibrant, urban Native groups, but people who would not really see a divide between the urban and the reservation. So that's how I grew up, migrating between the two, and seeing that more as a regular migratory route, versus the kind of divide that you hear a lot of reservation or urban people talk about in other parts of the country.

I was born to a mother who was always very political. My dad is white, but he was gone by the time I was three, so I was pretty much only raised with my Dakota family, although I did see my dad's family maybe once a year at Christmas. They only lived about twenty miles away, but they were not very involved. So I was raised by my grandmothers and my mom, going back and forth, and the American Indian movement was raging at the time. I started first grade in 1974, and my mom by that time had left my dad and she was an undergraduate. She went back to university with four children under the age of five. We lived in Aberdeen, South Dakota, where there's a Bureau of Indian Affairs office and a small college called Northern State College, which had a lot of Native students. My mom was pretty active with the American Indian student group there. Also, my brother's father is Floyd Westerman. The three of us girls have the same dad, but my brother has a different dad. And Floyd was active as a musician and an activist during that time. There were a lot of people like that in and out of our house—artists, activists, student activists. So I was raised to be very politically conscious from the time before I could read. I talk about this in the Introduction to my book, *Native American DNA*, that I knew that phrase *Custer Died for Your Sins*, from Vine Deloria Jr.'s book before I could read and I didn't understand what it meant. And so growing up like that, under the sort of political tutelage of my mom and then Vine Deloria Jr.—indirectly—really shaped my understanding that research and academic thought are political and were always part of the colonial project.

But my mom also emphasized that the only way out of poverty was through education for us. So we were also raised to think very pragmatically about education, that we had no choice but to graduate from high school. There was only a 50 percent chance back in South Dakota in those days that a Native person would graduate high school. But not only that, we were going to go to university. It was the only way that we could have a decent quality of life, and for us to give back.

SK: What was your route to academia?

KT: I never intended to become an academic, and I think most Native academics didn't. I actually interviewed Native American biological scientists and we all ended up on a research track kind of by accident. Like a lot of Native people, I intended to go to university and do something very practical. I thought I'd go to law school. My mom was a planner, without a degree, but she has always characterized herself as a tribal planner, or an "Indigenous planner" in the language we use now. She helped build a lot of community institutions. She was a grant writer for urban Indian survival schools in St. Paul and Minneapolis. She helped start our tribal school on my reservation, started the St. Paul American Indian low-income housing board, and helped found our tribe's drug and alcohol rehab center, which has a culturally based recovery curriculum. I grew up seeing her do research, so research for me was always about social change.

I never thought about being a professor. I had no role models for that in my life. I did an undergraduate degree in community planning at UMass-Boston and a master's degree in environmental planning at MIT. Then I worked as a planner for about ten years for U.S. federal agencies, for tribal governments, and for national tribal organizations. And it was good work, but I wasn't tremendously intellectually stimulated by it. I'm not a very good bureaucrat; I'm not a great grant writer. I see environmental policy as a really important thing and I'm glad I did it, but I didn't want to spend my whole career doing that kind of stuff.

It was by accident that I fell into a PhD. I was working for an Indigenous environmental research organization in Denver, an NGO. We had done a lot of work on tribal involvement in the cleanup and management of the nuclear weapons complex, and of course that's the Department of Energy's problem. But then the DoE just happened to start funding human genome research and had this grant opportunity and my organization started holding workshops to look at the implications for Indigenous peoples of the mapping of the human genome. And when tribal representatives and community people came to our workshops they had a lot of questions about the relationship between blood quantum, identity, DNA, and histories of colonial violations within research, and I was fascinated by those conversations, but I had no background in genetics. I realized I needed to go back and do a PhD, so I could write a book about the problems around the mapping of the human genome and genetic research on Native American bodies. So I did that. I went in with my dissertation already formed in my head.

I don't believe in doing a PhD because one wants to be an academic. I don't think this is a lifestyle choice. I did a PhD because I needed to think through a problem, and the only way I could think through that problem was to have four, five, or six years away from doing policy work to study and read and learn how to think in a more critical, complex way. Because you're not taught to think in those ways when you're doing policy work. And I question everything. I was somebody who just questioned all of the fundamental assumptions flying around me in that policy work.

I got through the PhD in four years, and as a result, the kinds of students I recruit are students who are older, who are coming back because there is a particular problem they want to solve, and a PhD can help them do that. Because this is not a good time to just decide to be an academic, right? It's a terrible job market; the university is being defunded by the public. It's actually a really dangerous economic and social space for people to be in, so I think one needs to be very, very careful about coming into this. So I ended up here totally by chance. And I thought I would go back out and work for an NGO again, but I realized that I'm a better academic—surprisingly—than I am a bureaucrat. Because I do science and technology stuff mixed with culture—a field that is much in demand and yet too few people are trained to do this, I'm doing really well in the academy, despite the fact that it's such a terrible place for a lot of people, which leaves me with no small measure of guilt. But it is what it is.

SK: Could you describe your relationship to the field of critical animal studies?

KT: It's by accident, right? When I went into the History of Consciousness [HisCon] at UC Santa Cruz, I went to work with Jim Clifford, and I didn't know anything about this world. I applied to two graduate programs. I applied to geography at the University of Edinburgh, because there was a woman there—Jan Penrose—who worked on questions of nationalism up in Nunavut. And I was very critical about the concept of Indigenous nationalism and I thought I could work that through with her, but I didn't end up getting any money. I also was accepted into the History of Consciousness program, which I knew nothing about. But I had been coming across Jim's work, *The Predicament of Culture*, in particular. I was always doing a lot of reading, even while I wasn't an academic. And I thought, well, I'll just apply where that guy is. And he and Donna Haraway accepted me as a co-advisee. When we opened my acceptance letter, which included an offer of full funding, my husband at the time, who's a geographer, said, "Oh my god, you're going to get to work with Jim Clifford and Donna Haraway!" I said, "Who is Donna Haraway?" And then I started reading her stuff and I thought, "I do not know what the hell this woman's talking about!" Anyway, I ended up working very much with both of them, and luckily they're very close colleagues and friends, and even though they work on really different projects, it was a very easy thing to be co-advised by them.

Because I was a student of Donna's I ended up being in class with so many people who were thinking about human–animal relations—all of the other Donna students. I always felt like I am the only one who's not a super nature person, and who doesn't have dogs, that kind of thing. Her students are like that. So I was sort of the provincial one in her group of students who was working on humans. I guess my interests developed in response to things my non-Native, animal studies, fellow graduate students were thinking about. We all shared something working under Donna, and that was an aversion to hierarchies and a desire to dismantle human–nonhuman hierarchies. Although I did not initially understand her language, Donna's ideas easily made sense to me. Everything that my fellow graduate students and Donna were talking about, in relationship to human–nonhuman relations, resonates with critiques that Indigenous people have made, even though I was raised in a family with no animals in the house, and it was kind of against our culture to, at least in my family, have them too close to you. They don't sit on your lap, and they don't sit on your furniture, and you don't touch them and sleep with them; it's kind of considered inappropriate. And you know how rez dogs are—they're kind of feral and running around living their own lives on the reservation. Their lives are hard because there's disease and they sometimes don't have enough to eat and all that. But I realized in thinking through this with these other students that I was raised implicitly, not explicitly, to understand that nonhumans have their own life trajectories. And I felt very averse to the ways that humans mess with nonhuman life paths, and that includes breeding them and making them too dependent on us. And, of course, I learned that there are long histories of human–dog companionship and relationships like that, and there's no disentangling their lives from our lives now. But there's something about that that seems really unethical to me. So intellectually, this stuff is really interesting to me, but I myself have never sought out these personal, intimate relationships with nonhumans; it's definitely been much more of an intellectual project.

Then I had a child—my daughter was born in 2002—and from the minute we sat her down on the floor in front of two big dogs she has been a complete dog person. I mean she sleeps with them, and kisses them; she's like Donna. In fact, she was debating with Donna over Facebook about some dog stuff. I reminded her, "Carmen, be respectful!" "I know you don't know who she is, but, you know, be nice. She's your elder." So I've got this child who brings all these animals into my life. I don't know if you want to write this, but I feel like I'm kind of like a racist—or a speciesist—in the original definition of "race" toward dogs because I think they have a right to exist but I don't want them in my intimate space. I don't want to touch them and be with them intimately. But I have to because of my child, and she cannot be happy if she doesn't have dogs in her life and her house. Part of my anti-speciesist education has been being surrounded by dog people.

SK: It's interesting that given the environment in HisCon, your reading of the multispecies literature, and your daughter's attachment to dogs, your relationship to companion animals seems not to have changed.

KT: No, I think they should have their own life. They have their own life ways, they have a social world, they have social practices, they transmit knowledge. They have their own intimate relationships among themselves and with some humans, and I think that's fascinating and that they have a right to that, and it's very sad how oppressive humans are in disrupting their social world. But I don't crave intimacy with nonhumans at all. It is very much an intellectual and theoretical project for me. I think a lot more about filling in some of the gaps in the new materialisms and critical animal studies. I mean the whole idea of an animal, right? In our stories, in our languages we talk about all living things' being in relationship to us, but living things aren't limited to the things people in the West think are living, because spirits are in relation to us, rocks are in relation to us, things that people in the West think are dead are alive and they're related.

So I prefer to think more in terms of who we're related to, but it doesn't necessarily mean that that translates into some sort of intimate companion relationship for me; it's more the way an ecologist might think about being related. That we're interdependent, that we have intimate histories, that maybe we have moved apart as communities of beings. I don't really see a lot of difference between evolutionary theory and the way that Native people think. I think we have a sense that there are long histories of relatedness and there have been close relationships that have gotten further apart over time, as different communities of beings drift apart.

SK: Moving now to the connections between your philosophy of relatedness and practices of eating, could you describe your approach to consuming meat or other animal products?

KT: For me there's no good way to consume anything that we put in our body in the kind of society that we're living in. I have this conversation a lot. I was actually dating a vegan until I left Austin for Edmonton. I don't know if all vegans are like this, but I would call him a very typical vegan where he's got this sense that he can stand in this morally pure place because he's a vegan. Yet he shops at Whole Foods, and he buys fruits and vegetables that are being shipped using fossil fuels, and his soy is being produced in the Amazon and that's displacing humans and nonhumans. The problem is our food system, but that doesn't diminish the cruelty. There was a *New York Times* article that came out about an agricultural research station in Nebraska just a couple of months ago; it was just horrifying the way that these animal bodies are treated. It's torture what they're doing to them in the course of research and then production of meat. For me the question is our whole system of producing and consuming. All of the bodies that are used to prop up that system, whether they're human or nonhuman, are being violated and exploited.

I teach a book called *Hunters and Bureaucrats* by Paul Nadasdy that's about Yukon First Nations hunting people. If you can hunt, I think that's an ethical way to consume meat. They have these sets of relations with their prey. There's an acknowledgment that humans in their culture have also been prey. Most of us obviously don't consume meat like that, but most of us don't consume vegetables in a proper way, either.

SK: Has your thinking about food and the food system changed over the course of your academic career, or throughout your life?

KT: Growing up on the reservation, you ate what foods you could afford. And a lot of that was, we called it "commods"—commodities—U.S. agricultural surplus that they distributed on reservations and to other poor people. I'm trying to think when I first started thinking about meat consumption. It was probably when I moved to Boston to go to university in 1989, because I stopped eating meat for a while. I don't remember why I did that, though. I don't think it was an ethical thing. Maybe I had some vague sense that forgoing meat was healthy. I don't know. But certainly, growing up, both in reservation and urban Native communities, nobody was a vegetarian, nobody talked about that, it would have been considered bourgeois and urban, or something white people did. But the other thing is, I grew up in a hunting family. My grandfather hunted, my uncle hunted. My grandfather, grandmother, and uncle fished. So we had a freezer in our basement that was always filled with river and lake fish, elk, and venison. We had dried venison as a snack a lot. A lot of our meat we caught ourselves, but not all of it. We'd get hamburger from the store if my grandma wanted to make spaghetti, but then my grandfather would leave the house and go to the café because he didn't consider that real food. I would probably say at least half to three-quarters of our meat was caught ourselves.

That said, I wasn't a big meat eater. I remember refusing to eat a piece of meat when I was a kid because it had veins in it. I thought that was really disgusting. So maybe I was always somebody who just thought eating flesh wasn't really that appetizing. And the other thing about my grandfather and my uncle—my grandfather was a white guy, because that was my grandma's fourth husband, while my uncle and my grandma were Native—they didn't hang trophies on their walls, they didn't believe in stuff life that. My uncle says that's disrespectful to that body, and then he also stopped using guns and he went back to hunting with bows and arrows. But my uncle's a national archery champion, so he would not advocate that just anybody do that because it would be cruel if you weren't good at it, if you didn't kill with one shot. So I did grow up in a family that kind of thought about these things, and there was no sport hunting, ever. We ate everything, or gave it to other people. So I suppose there was a foundation there for me to think about these things in more complex ways, because we had a closer relationship to our food—at least growing up. But then there was all the commod stuff too, that wasn't close. And when I was a kid we still had gardens on the reservation. You don't see that now, but my grandparents had gardens.

SK: Do you have a story you could share about a time when your dietary choices or food philosophy has been the subject of conflict, pleasure, or awkwardness?

KT: When I moved to Boston and I stopped eating meat for a while, I would come home and my uncle, the one who hunts, would tease me mercilessly, and he still does. Because I'll eat tofu, I have some good tofu recipes, and because I have so many vegetarian friends, and I'll get teased about it, because that's how it is in small towns. I also got a wheat allergy diagnosis about four years ago when I was living in Berkeley. I remember being so annoyed with people in Berkeley. Nobody could eat anything. Everybody had all kinds of food rules there, and they judge you, they really judge you in Berkeley about what you eat; it's kind of oppressive. It's a Michael Pollan world there. So I would be busy making fun of these Berkeley types, and here I get a wheat allergy. My hair was falling out. I was getting terrible rashes on very specific parts of my body, and I went to the allergist and she told me what I had. My hair had been thinning for years, so clearly I'd had a problem for some time; I ate too much wheat. So I had to go on the wagon with gluten, and it's like you're a pain in the ass to people. I'm embarrassed to say I have this allergy. But I was really strict for three years and now that it's really out of my system I find that I can cheat sometimes, and I'm relieved about that.

SK: Given the systemic problems with the food system that you've identified and the political limitations of responding to these problems in an individualist, consumerist way, I'm wondering how you approach shopping or otherwise procuring food.

KT: Being an academic with a middle-class income, I can buy organic. But that's about my own health, knowing that I don't want to consume pesticides and I don't want to increase my own risk of cancer. I don't know that it's really a social justice thing. I try not to shop at stores that I know have really bad labor practices. I can afford to buy what is labeled as free-range meat and things like that, although my vegan friend in Texas would always post to Facebook these articles saying, "That's not true, they call it free-range but it's a still really violent part of the food system." Maybe that's true. These labels facilitate people's shopping choices. Do they facilitate a real change in agricultural practices? I do try to consume less meat and fewer things where I don't know anything about the history. But you're reduced to looking at labels. Is it free-range? Is it free trade? And even though I'm this middle-class, conscious person now, I'm sure I ate in a much more sustainable, humane way when I was a kid, because we procured so much of our food from our own labor. We kids even "hunted" nightcrawlers on nights after a rain. Our grandparents used them for fishing. And there was no moral project there, it was what my grandparents could afford, it was the traditions of people—it wasn't just Native people—the white people where we lived, too, they all hunt, the boys all get a BB gun when they're thirteen, people all had gardens. It wasn't any big social justice project; it was just what people could afford and how they lived.

SK: This project was inspired in part by an interview with Donna Haraway in which she said she thinks there should be a place in the world for agricultural animals, but that when she's confronted by the convictions of her vegan friends, she doesn't feel like she has adequate answers. I'm wondering how you would respond to her response.

KT: Because I've been around Donna so much, I actually do feel like her response to ethical veganism is a really good response. The thing I noticed about vegan friends of mine—they're still basically operating under the same hierarchy in which humans are above nonhumans. And instead of being able to exploit those nonhumans at will, we should steward them and take care of them. That is just not the worldview that I operate under. It's easier for them to care about a nonhuman that they have judged to be closer to a human, in terms of brain size or transmission of culture, or whatever. They're not thinking about relations as much as they're thinking about the ethical job of superior humans to steward the lives of those that are less than they are. I feel like Donna gets this, and

I've seen her take to task vegan scholars at conferences. Actually, we were at a meeting in May 2013 at Berkeley that my friend Harlan Weaver organized. Harlan's just fantastic. So Harlan organized this meeting at Berkeley and Donna was there. And there was this vegan guy from somewhere in southern California. Just a totally urban guy living in L.A., probably does all his shopping at Whole Foods, giving a talk on veganism and being so judgmental. And I'm sitting there as an Indigenous scholar thinking, "What the hell! You can go to Whole Foods and live your vegan lifestyle, but you're going to tell this to hunting people in the North? Just no cognizance that not everybody can make those kinds of choices and live like that." But I wasn't going to say anything because I was feeling really angry and that's not productive, but Donna intervened and she made a comment, "So what are we going to do if we end the agricultural food system? You know what that's going to result in? It's going to result in the extinction of all these animals bred for human consumption." And this person was like, "Yeah," and he admitted that he thought their extinction was acceptable. It was really interesting, the lack of thought about that. And I haven't really probed her on her thinking about the role of those kinds of beings in the world, but there are these discourses of purity that are underlying these kinds of ethical approaches to food, and I just don't think there's room for that kind of purity in this world. It seems very Christian to me in some ways.

SK: It's interesting that you should say this, because two of our other interviewees have described their conversion to veganism as a spiritual experience of sorts, something that just came to them, that they had no choice but to undertake.

KT: So I met Chloe Taylor at a party here in Edmonton when I first arrived and she's a vegan, and I told her, because I had been cooking for this vegan boyfriend of mine, "I have an amazing vegan sauce recipe, it's really good, it tastes like meat." And so we got to talking and I said that I see us in relation to our prey and I come from a culture in which not too long ago, we were prey, and hunting people still can be if they're out on the land. A lot of human beings in this society haven't experienced what that's like. But I grew up in a way in which nature was dangerous. That's why I don't hike and do all of this nature girl stuff. To me it was life and death: Blizzards can kill you, tornadoes can kill you, floods can kill you. That's the kind of landscape I grew up in. It's beautiful and violent and dangerous. And so we are not superior to nonhumans, we are at their mercy to some degree, and this is why we had guns and we had other things to protect us when we were out there. But I also told Chloe, because I view myself as being in relation, when I take that to the ultimate edge of thinking, I actually come to see now how certain peoples practice cannibalism. Because if we live in relation to nonhumans, it also makes sense that we should be able to eat our own if there was nothing so elevated and special about us. I get to that point in some of my classes sometimes and my students are horrified, but why is that such a horrible thing? I'm not saying you should go out and kill people for food, although then one could say you shouldn't go out and kill nonhumans for food either. But if we are capable of being the prey of nonhumans, I think ultimately, morally, we're capable of being eaten by each other. Chloe Taylor said she agreed. She said, "And I don't want to eat humans either!" And we laughed. But for me, if I'm going to eat meat, I can imagine a circumstance in which if somebody offered me some human meat and it was willfully given, it would be disrespectful to turn it down. So it was a very interesting conversation we had.

But it sounds kind of crazy to most people because there's such a deep core assumption in our society that humans are so special and that somehow cannibalism is just an incredible, terrible sin.

SK: Are you working on a specific project in animal studies at present?

KT: No. My next book is called *Pipestone Relations*, and it's about the Pipestone quarries. A lot of what I'm doing is looking at the absence of Indigenous thought in the new materialisms literature and in critical animal studies. There is a lack of understanding that Indigenous people have thought about questions that are central to these fields. Indigenous thought is largely missing from these academic conversations, and I think that Indigenous people should be much more a part of these conversations and should be at the forefront of theorizing these things. Our language in "academese" and in English for talking about these issues is inadequate. First of all, there's not an animal–human divide for many Indigenous people. And that is not to say there is some utopian idea like "We're all equal," but it's just a recognition that, why would you have the word *animal* when we're all related? The word *animal*—everybody in animal studies knows this—it's used to denigrate, to de-animate, to make some beings less than others and controllable by us and exploitable by us. And that word gets applied to both nonhumans and to humans. And it's really anti-relational; it's a horrible word. And even if you don't care about nonhumans, people should care about the question of the use of the category of animal because it's used to enslave and incarcerate and kill a lot of human beings. And even if that's all you care about, that's enough, right?

SK: I have two questions related to this point: Do you see yourself in conversation with posthumanism? Do you have thoughts about posthumanism as a category of analysis, or as a way of thinking about the human?

KT: To speak broadly, a lot of Indigenous thinkers already talk more about "persons." Humans and nonhumans can be persons, and so can "spirits," for lack of a better word. And we think about collectivities, which you might translate roughly into English as *peoples*, or *nations*, but those are bad translations. This is part of the work of getting Indigenous thought into other academic conversations; there are much more complex ideas within Indigenous languages that really get lost in the translation to English. When you dig into those languages, it just floors you, to find the deep and very different philosophical assumptions. There are very different worldviews and categories embedded in those languages that give us a whole new way of talking about these sets of relations. And I think that's some of the work that needs to happen. We need to be working with those who are revitalizing Indigenous languages and thinking out of those languages. So, I don't think Indigenous thinkers would spend a whole lot of time trying to critique or save the concept of the human. I think that there are other terms, if we were to get deeply into this conversation, that we could use.

SK: Is there a key concept or theoretical perspective that guides your thinking about human and nonhuman animals or other life forms?

KT: Yes—"being in relation." I was watching a film for my Pipestone project where Albert White Hat, who was Sicangu Lakota, and he's a language speaker, was talking about Pipestone and our relationship to the stone and the pipe. A lot of people talk about that site as sacred, and the stone as sacred, and there's an origin story that lends credence to this notion of the "sacred," but White Hat didn't use that word. And the word that we tend to talk about as being sacred that people translate from Lakota or Dakota into English as *sacred*, he didn't describe it that way. He translated that word as *relations*. He didn't say *relationality*, but that's what he was getting at. I thought, "Holy, Albert just said this word that we use as *sacred* really means *in relation*." We have inherited so much from Christianity that it has completely perverted our ways of thinking about how we should relate to these things. And that's what my book is about—taking the concept of relations and getting away from the concept of the sacred, which is going to be controversial. There's another Native studies thinker who is doing a lot of work along this front, David Shorter at UCLA. Because, again, you can't have the sacred without having the profane, but we don't have anything profane in that land. There are some things that are more revered than others, like your elders are more revered than a younger person, or some mountain might be more revered, but it doesn't mean that the other stuff is profane. Again, these kinds of binaries that exist in English and in histories of Christianity are really counterproductive.

And so that book is going to talk about the relations between humans and rock, between the U.S. Park Service, federally recognized tribes, and another organization called the Indian Shrine Association. It's going to think about the relationships between quarries and carvers, between climate scientists and ecologists, and between the Park Service and the tribes. So there are all these aspects of the Pipestone quarries that you could consider "religion," you could consider them science, you could consider them policy, and I want to show how each of those categories is problematic and instead think about the relations between the humans, the institutions, and the nonhumans, including the stones. So—*relations*—that's my term right now. It's very much in conversation what's happening in the academy, which is why I think this is an ideal time to be doing this work. The academy is getting to the place where we can actually have a conversation in a way that probably wasn't possible twenty years ago.

CHAPTER 4

The Tyranny of Consistency

Naisargi Dave

Anthropologist Naisargi Dave credits an undergraduate feminist ethnography class for introducing her to the promise of her discipline. The child of Indian immigrants, Dave was born and raised in Atlanta, Georgia, in the late 1970s in a "rather culturally conservative" household wherein she began to question the governing power of cultural norms. In her interview Dave shares her thoughts on hypocrisy and how what she calls the "tyranny of consistency" shapes attempts to live ethically. Last, with pluck and candor, Dave shares the story of how she became a vegan and thus helps illustrate her provocative observations about how and why practice can and should precede ideology. We interviewed Dave in Toronto on October 9, 2015.

VICTORIA MILLIOUS: Because we are interested in how personal and academic interests are mutually constitutive, let's begin with biography. Could you talk about when and where you were born and raised and your formative cultural, intellectual, and political experiences?

NAISARGI DAVE: So I was born and raised in Atlanta, Georgia. My family is from India, from Ahmedabad, Gujarat, specifically. And they migrated to the United States in the early '70s. I was raised in a suburb of Atlanta, known as Stone Mountain, one of the few Indian students in my classes, and for reasons that I'm only now starting to really think about, I wasn't very attuned to my racial difference.

My family was—as many Indian immigrant families tend to be—rather culturally conservative. Their thinking was that we have brought you to this strange land, and you're vulnerable to all of these other kinds of influences, and we need to protect you from that so that you can be raised with the same values that we were raised with. So as you are encountering other people, you see that the things that are so normal for you, the rules that are normal in your household, don't necessarily apply to other people. The question of hair was a formative one for me and tends to be for queer people and queer women in particular. I had to always have long hair, and my mother had to have long hair, and I would ask, "Why can't I cut my hair?" And it always came back to, "Well, because that's our culture." So I grew up very intrigued with and angry about this concept, that it closed down every question I had. Every curiosity that I had about the world was just met with this one word, *culture*, as if everyone knew what it was and everyone agreed that it was important and relevant and could govern my life.

I was pre-med and grew up with a very functionalist sense of education. Education is that thing you do so you get a job and you make money, and you make your parents proud, and that kind of thing. But I went away to college and encountered an anthropology class as an elective, and I fell in love with that course—it was a feminist ethnography class, actually. It was the first time in all these years that I could actually discuss the culture concept. This course took up culture as an object of interrogation and not as an a priori good that everyone understands and agrees with.

In terms of formative intellectual experiences, I'm struck by how important friendship was to me. That was my entrée into other worlds outside of my own. The most important lessons that I learned and that I can still see being formative for me, intellectually, are ones that I learned from my first good friends. In high school, it was through friendship that I started thinking about intellectual life as creative and not just functional. It was in friendship that I learned how to debate, question, and argue, instead of the mode of simply asking a question and receiving a supposedly "correct" answer.

VM: And in terms of your moving through to your graduate student experience, could you speak to that?

ND: I took a women's studies course in college, and that was where I decided to become an academic, maybe before the feminist ethnography course. My women's studies professor, Nina Karpf, at the University of Georgia, was a very important figure in my life. She was political about the fact that she didn't have kids, which was also radical to me. It was important for her, and she made a point to say this all the time, "I do not have children, because this is my life. My life is teaching, and you come into my life, and you leave my life, and I'm okay with that. Every year is a different emotional experience for me." I found that amazing. I find myself thinking about her a lot now, too, in terms of how I organize my own life. But I joke about this as my first "coming out" to my parents, which was to tell them I'm not going to be a doctor, and I'm going to go to graduate school. I had read Ruth Behar, Catherine Lutz, and Kamala Visweswaran in my feminist ethnography course, so I applied to the schools where they taught and wound up at Michigan, which was wonderful. My advisor was Jennifer Robertson, who works on sexuality and technology in Japan. I admired her profoundly. So much actually is generated in admiration, in looking up to somebody as a model of how to organize one's life and how to live well, and how to live a creative, interesting life.

VM: So, thinking specifically about animals, what do you see as the primary purpose of your academic work, and what are your motivations for engaging in it?

ND: I honestly would say that the primary purpose of my academic work is pleasure.

VM: That's what we needed to hear today!

ND: I'm not a terribly artistic person, so this is my mode of expression.

VM: Your craft.

ND: There's nothing else I would rather do, having the freedom of coming up with ideas, expressing them. That, to me, is the main purpose. Though I confess that I feel more of a political purpose with my current project. With my book on queer politics, for example, I didn't aim for people to become queer when they read my work. That was not exactly the object, neither of my book nor of my teaching. But with this work on animals, I like the thought that someone might think differently about their relationship to animals after reading my work. I wouldn't frame that as a motivation, though, because those sorts of motivations fail. When you try too hard toward a communicative goal, that effort limits expression, and the things in the field that I'm influenced by are fundamentally expressive. It's less for me about trying to communicate a specific idea and more about ethnographically conveying the affective experience of being caught up in something.

VM: Could you explain how you came to write about animals?

ND: Sure, I'll admit this: I went to India often as a kid. And that Western liberal sentimentality, I experienced that myself: the sight of animals on the street. I think those childhood experiences were formative. I still remember the very specific encounters with specific animals.

VM: Can you give us an example?

ND: One was in Ahmedabad. I was out near my mother's home, where she'd grown up, and there was a donkey, a laboring donkey—it was just a hot, hot day—and the donkey was tied by a short rope on a busy street. There are cars and trucks and people swarming around, and the donkey was saddled with these bags of concrete. Its head was bowed from the weight, and it was just staring at a pole—and there was something about its loneliness, as I perceived it. The fact that this being didn't experience any beauty or pleasure in life and was standing there saddled with concrete. It made me think a lot about hypocrisy, actually.

There is an assumption about India in the West that we're all vegetarians and we worship cows and are good to animals. But vegetarianism is specific to Hindus and, specifically, to high caste Hindus. I grew up in a teetotalling vegetarian household, and my mother's neighborhood in Ahmedabad is a very Brahminical one. There is a widespread belief that people are kind to animals because they don't eat them. But there were examples everywhere of the complete destitution of animals, the refusal to acknowledge their pain or their suffering when they're in your midst. And this did, for me, raise the question of hypocrisy, if we can put it that way. Though, as I'll get back to later, I've been rethinking hypocrisy and our critical relationship to it.

Another thing that felt very specifically Indian to me in relation to animals happened in a human rights NGO, populated with young progressives. A foreign visitor, during lunch, asked for the vegetarian dishes. And someone turned to her and said, "I'm sorry, but we are not vegetarians here, vegetarianism is a tool of the right wing. It's coercive, and it's oppressive, and it's elitist dietary politics." Things of that nature. This seemed such a very different context from a Western one or, really, anywhere else. I would be hard-pressed to think of a place where being progressive precisely means militantly eating animals, whether you want to or not. I know Indians who grew up vegetarian and they eat meat deliberately to, as they say, reject their caste privilege, to prove this point about their solidarity. So that to me was a very interesting phenomenon. All of these things together are what led me to start writing about animals in India.

VM: In an essay in *Cultural Anthropology*, you discuss how meat eating is often justified as a form of resistance to Hindu nationalism and religious and caste chauvinism in cosmopolitan and progressive circles. This analysis is compelling for what it suggests about the power of ideologies we vehemently oppose to colonize and discipline us. Can you talk about this observation and how, as a scholar of activism, you have come to approach big questions like "social change" and "resistance"?

ND: I love the way you put that: how the ideologies we vehemently oppose colonize and discipline us. There was an event recently that exemplifies this. And dozens of people have forwarded me news stories about it lately, and I feel like the unspoken implication is, "See what you're participating in as a vegan in India!" What happened is that a Muslim man was accused by Hindu villagers of having killed a calf and eaten it. Based on this rumor, a mob gathered and killed this man, beat one of his sons to within an inch of his life, beat his wife or his mother, and beat another one of his children. This is an all too common example of the way the Hindu right mobilizes the cow—not the animal in general, but the cow—as a symbol to enact violence on other communities, whether they are Dalits, or Muslims, or Christians. This is a thing that happens. So accepting that, and acknowledging the way in which vegetarianism plays into a politics that we abhor: What do we do with that fact? That's an important question for me. All too often, what we do with that fact is to say that to oppose that kind of violence means to eat cows. It's posited as a kind of solidarity with non-caste Hindus. But of course the Brahmin who decides, "I'm going to eat beef—or pork or chicken or whatever—" will simply not be subject to the same kind of violence as someone who eats meat by tradition. What this high caste person doesn't understand is that people aren't subject to violence because they eat beef: They are subject to violence because they are Muslims or Dalits. The cow was the mobilizer, it was the justification, but it was not the reason. So when a Brahmin decides that they're going to start eating beef, no matter how much they try to be in solidarity with the Dalit or the Muslim, it's not going to happen, because what they fail to recognize is they're just playing into the circulation of symbols that is at the heart of the problem in the first place.

One comparative case is the right wing in the United States, which at times enacts violence against women who have abortions, or the doctors who perform them. We as feminists vehemently oppose that, yes. That does not mean, however, that we go out and deliberately have abortions. Now, they're not one-to-one comparisons, of course, but the one reason they're not is that we value humans and animals very differently. Of course we don't go around having abortions or aborting other people's fetuses just because we oppose a Christian right ideology. But the issue there, then, is that in understanding the animal as a symbol that is not subject to the same value as a human, you're participating in and allowing your own ethics to be completely colonized by the use of symbols of an ideology that you oppose. That's not a rejection of an ideology, it's an adoption of an ideology, it's speaking in those people's terms, it's using their morality to be the only framework in which you can possibly be a moral person. Your only option then is, do you eat beef or do you not eat beef? That's not rejecting the use of the cow as a violent symbol; it's appropriating it, and not appropriating it in a way that undoes the problem. It's more of a wholesale adoption of an ideology than an appropriation, actually.

You also asked about resistance and social change, and I think it's important, too, to not see every incidence of this sort of adoption as closure. This is one of the things I was working out in my first book, as well, that we often tend to have very dramatic ideas about social change. It's important to recognize that limits and problems actually provide the impetus for new ways of thinking, rather than simply being an obstacle or a problem. So, in this case, the moral universe in which I can think about my relationship with animals in India is more or less closed, colonized by the Hindu right, for example. That's our closure. But what does that closure enable? What emergences are possible there as responses to the limits, which then become foreclosed themselves in different ways, or normalized, and on it goes? I think it's important to think of social transformation more cyclically, and to acknowledge that everything that we see as problematic is a possibility, an opportunity, and to think more about closure and immanence.

SCOTT CAREY: Could you talk about a key concept or idea that guides your thinking about humans and animals? Earlier, you mentioned hypocrisy as something you'd like to discuss.

ND: Exceptionalism is one. What makes me think that I am so special, that for my comfort or my pleasure I can take another being's life? Granted, I tend to buy and live vegan, so I'm not actually faced with a lot of dilemmas, but when I am, this is the question I ask myself. Another question, inspired by Montaigne, is this: "What do I know?" I feel at my most open and creative when I imagine myself as quite dumb, as not knowing much of anything at all. I don't know if a pig suffers, and I don't know if it doesn't. I don't know that a donkey feels lonely and I don't know if it doesn't. How on earth would I know? But it's accepting that I don't know, rather than insisting that I do, that allows me to be open to different imaginative possibilities rather than wasting my energies defending my ostensible certainties.

And on hypocrisy—thanks for bringing that up. Originally, I used hypocrisy as a negative. Who are these people who go on and on about how they respect animals and then leave a donkey tied to a pole, or throw a cow in a garbage heap? But I started rethinking my relationship to hypocrisy, or to consistency, in the last few years. Here's why. A student came into my office hours and started crying because she wanted to become a vegetarian and her family and friends responded by pointing out how futile it was by showing all of her contradictions: "What is your bag made of, what are your shoes made of, don't you drive a car which requires roads which destroy forests and the animals who live in them, and so on?" These aren't illegitimate questions; they're interesting food for thought. But I've started to think of the problem of hypocrisy also through the tyranny of consistency, which is related very much to identity politics and the recognition that people who do things normatively, who eat anything, who are heterosexual, whatever the case, they do not have to explain the hypocrisies in their own life. It's only the person who tries to do something different who is then subject to the problem of contradiction. Contradiction itself has become very important to me, and in India, from an ethnographic perspective, it's important. People ask: "Isn't there a contradiction in the fact that you care about animals in a way that coincides with what the Hindu right thinks about cows or animal protection?" But also these everyday questions of what does it mean to practice this and fail to practice in this way. So the imperative of perfection, the idea of consistency and how important that is, always doing things the right way, which of course, even people who do things normatively don't do, but they're not held to that standard. They're not asked to account for themselves in the same way. This made me reflect a lot more on contradiction as a problem.

That contradiction doesn't actually exist between two things but exists in the frame, in our concepts of things, not in the things themselves. It's only when we frame two things as comparable in the first place, and as the bounded comparison, that two things can appear as contradictions. Once one opens one's scope, or recognizes that one's concept of this thing—animal, meat, purse, leather, etc.—is actually more flexible than we think, the contradiction disappears. So contradiction is a problem politically, in terms of the imperative of consistency that forces anyone who tries to do anything differently to make a choice between nothing and everything: I either do it perfectly or I might as well just do nothing. That, of course, is a terrible way to operate, but it tends to be the way we approach ethical life. Except for all the invisible ways in which we're never perfect moral beings, but we excuse them for various reasons, because they're normative and thus naturalized. The accusation of hypocrisy is a problem politically, but I think it's also an interesting problem philosophically, and so, in my new book, that's one of the things that I'm trying to think through.

SC: You emphasize in another interview the importance of doing intersectional work on animals. This is a common battle cry within critical animal studies, but one that is difficult to put into practice. Can you discuss how you have practiced intersectionality in your work on animals and humans in India and perhaps why critical animal studies has struggled so much to overcome its limitations in this regard? How might this be related to your other claim, that animal activists might not be as self-reflexive and self-critical as other activists?

ND: Well, I can start with the latter, and I feel okay saying this because I had this conversation with a very self-reflexive animal activist. In a way it circles back to what we were talking about earlier—how the starting ground is so different when you're talking about animals and you're talking about human rights, that it's very difficult to think about intersectionality when you're so completely dismissed by other social movements. I guess it was slightly different in the case of queer activists in India, who are of course, for the same reasons, seen as bourgeois, elitist, having concerns that had nothing to do with real Indians and all of that. Queer activists were looked down upon by more established movements, but the thing is, everybody also realized that there were lesbians and gay men in all of these organizations already. I think that the lack of a certain penetration of animal activists in social movements is part of the issue; but for me it also comes down to the problem of emergency. Where—and this was a problem in queer politics as well—there is so much going on all the time for an animal activist. There is always an emergency. You walk down the street and there are ten things that you could put your attention on, and that make a demand on you. And it does feel like some sort of luxury; it is decadent to spend our time in a reading group, in a room reading Derrida, or Peter Singer, even, when there are things that one could actually be doing. Now this itself is an interesting kind of distinction between the work of contemplation and hands-on labor, one that's also very divisive within animal rights activism, at least in India. But the people who actually go around working in shelters and picking animals up off the streets are thought of differently than the ones who do more of the organizational kind of work. That to me is one of the issues around intersectionality; it's just the urgency and the immediacy of things that need to be attended to, and intersectional work seeming like a kind of decadence: Why bother trying to persuade people about this cause when there are things that I could be doing instead that would be more useful?

SC: Could you explain to us how you became a vegan and your current relationship to consuming meat or other animal products?

ND: This is a story I didn't tell my students for the longest time, and then I decided I would just admit to. Because there's a lesson here. I became a vegan to impress a hot woman when I was in college. I lived in a rented house and there was a person who worked for my landlord who was mowing the lawn one day. She had great muscles, she looked good. I asked, "Who is that?" I was told her name, and the next thing I was told is "She's a vegan." This was very unusual back then. But because I thought maybe it would be impressive to her—she who had never even met me—I just decided one day when we were all at the Waffle House to be vegan. I ordered dry toast and black coffee, and that was that.

I grew up vegetarian, but then did what many vegetarian Indian kids do, which is to rebel and eat meat in college. So I went straight from meat to the dry toast and black coffee at the Waffle House, and I've been vegan since then. That was it.

The reason I became okay with telling that story to my students is that in that moment, I didn't know the first thing about veganism. I hadn't worked out why I should do it, or what the politics were. It was just a one-off decision made on a lark, kind of as a joke to my friends that day, that then ended up becoming one of the most important things in my life, politically, ethically, intellectually. I think it's a useful story because it demonstrates that the practice can, perhaps even ought to, precede the ideology. So long as we are practicing something, it's very difficult for us to see outside of that thing because we're so invested in that practice, in maintaining ourselves and our lives as they are. It was only when I was no longer eating meat that I was free to think about what meat was.

So long as you try to fix the ideology before the practice—you can say, "Well, first I have to convince myself that veganism is worth doing"—it's never going to happen because there's no way to ideologically explain it, because everything is geared toward maintaining the norm. There will always be the one person pointing out the hypocrisy, the flaw in the logic. So if you wait, and wait, and wait to have it all figured out, then you're not going to get there, and that's not accidental, that's by design, that's how normativity reproduces itself.

I've remained a vegan. I was a primarily dietary vegan for a long time. I think it was when I started doing my research for this project, and seeing a calf tied by a short rope to the ground, where it would stay for its whole life, providing milk, that made my veganism more militant, perhaps, more holistic. Now it applies to clothes, household goods, products, everything.

SC: Have your thoughts and feelings regarding the consumption of animal products changed over the course of your scholarly career? We were wondering, in particular, if you could elaborate on a point you made in another interview, where you said, "Had I to decide now, I doubt I would be a strict vegan."

ND: Let me explain what I mean since I left it deliberately vague there. What I meant is that my practice would remain the same. I'm not just going to suddenly start doing things that I find personally appalling, or gross, or problematic. That's not what I mean. All I mean is that I now recognize the role that the tyranny of consistency plays, not in furthering what I consider to be ethical goods but in preventing them. What I mean by "had I to decide now, I doubt I would be a strict vegan," I just mean that I wouldn't identify as such, I would just live the way I live without the label. Because the contradictions only arise when you create a thing that can then be compared to another thing, and some contradiction can be recognized there. I also feel like it's politically not efficacious to make something that I feel strongly about be tied in with my identity. The fact that somebody thinks that they have to "become a vegan" in order to change their relationship to food is, I think, terribly problematic.

Let's create practices in which contradiction isn't even possible, in which it's just an ethical life, but not bound by new norms and new rules that then have to be overcome, or transcended, or questioned, or interrogated necessarily. It's also not efficacious to tie ethics to politics because, in a way, and this is a slightly more abstract reason, it's kind of daring somebody, it's a power play. If I convince someone to become a vegan, they've essentially subjected themselves. I say, "This is my ethics," and for you to agree with me means for you to become like me. Who wants to do that? I know I reject anyone telling me what to be, and what to do, and what to think, so of course anybody else does too, and why would somebody want to subject themselves to somebody else's moral regime? Also, just from a perspective of what's efficacious and what's not, I don't always think that veganism, as an identity category, is all that persuasive.

SK: When I read your work, I think there's no way I could have done that fieldwork, witnessed that torment, which is probably not the effect—or affect—that you're trying to induce in your reader. It made me think about the limits of my own ability to experience animal suffering up close.

ND: Well, it was a huge revelation for me that I could, because people send me awful videos all the time and I don't watch them, nor do I watch documentaries about the meat industry. I don't open these links, I don't subject myself to these things. What I realized in the field is that the difference is between more and less mediated experiences. There was something about the immediacy of these situations, of being in a slaughterhouse, of literally standing in blood. There are several things going on here. One is just a simple numbing effect. You're overstimulated; you can't cognitively bear everything that's going on around you. I would slip into something that was almost like a—and this isn't something I share, and so I might later regret this—but there was almost like a giddiness, a euphoric feeling.

That to me was really disruptive, because it made me realize how easy it could be to normalize killing, even to kill. I talk about this in another paper that just came out on the poultry industry, and there's a moment where I'm watching these birds being unloaded from a truck, and it is a horrible scene. But I had a moment where the guys who worked in the factory were watching me with amusement. Here is this Western short-haired chick who's probably freaking out right now. And what I wanted to say to them was, I feel exactly the way you do about these birds. It is really hard to think about them as anything other than objects, when they are placed this way. You get into a mode of "I'm alive, and you're not." It's amazing how—and I talk about this in the poultry paper—a twenty-year vegan, bleeding heart, I cry at the sight of a lonely donkey, and yet here I am, suddenly just heartless. Just utterly heartless.

SK: Is there anything you want to add about the relationship between your academic work and your personal practices?

ND: Well, in a strange way writing this book has made me less of a vegan in identity terms, more of a vegan in practice, if that makes any sense [laughs].

SK: Such feelings are hard to shake, but do you think you've become less judgmental of people's attitudes toward eating animals as a result of the problem around consistency and hypocrisy that you've so eloquently explained [laughs]?

ND: Actually, no [laughs]. I wish, I wish. As you say, we can't escape it. We're judgey about all sorts of things we feel strongly about. I think it has changed my relationship with humans, but I'm not exactly sure how. Even in my queer life, I can sense that same movement where I think I'm more queer in practice, and much less queer in identity now. Being less the label enables me to be more the practice. I'm less constrained by the rules, the proper way to be this or be that and how to present myself.

I definitely think that the veganism and my political relationship to animals play a big role in what I write. And I do want to separate, as much as it might be difficult to separate them, the question of being motivated by a certain message that I want to get out, because I don't know, actually, what that message is. I absolutely have no idea. I think the objective is just telling a good story, not good as in entertaining, but something that affects. And I have no idea what the effect is or is not going to be, but if you don't affect, then there's no question about what the effect is going to be or not be—because there isn't one. Writing about animals has made me more queer in my writing, because I know that having a message backfires. I don't seek to have a message; I think adamant messaging is part of our moral problem. So it's that I want to write truly and queerly.

SK: Can you tell us a story about a time when your veganism has been the subject of awkwardness, celebration, hostility, or something else entirely?

ND: Well, there's a story about my mom [laughs]. My mom was actually more devastated about my being vegan than about my being a lesbian [laughs]. I remember one day she made something, and this was very early on in my veganism, and she hadn't quite picked up on my rules of what I could and couldn't eat. And she worked really hard on this dish, *shrikand*, something I really liked as a child. And I, of course, with all of my ideological righteousness, said, "I'm not going to eat that!" And she asked, "Can you explain this to me?" And I said something about cruelty to animals, and she said "Cruelty to animals? What about cruelty to mother?"

[All laugh]

ND: Her whole position, which is one I hear often, and I think Jonathan Safran Foer raises critically in *Eating Animals*, but something Michael Pollan advocates, is that people's animal politics make them cruel to people, because it doesn't allow them to break bread with them. I think that's absurd for the reasons that Safran Foer talks about: What it means to break bread is precisely to talk, to communicate. My simply going along with what everyone else around me is doing is not something to be valorized. Commensality is discourse. Being normal isn't conversation.

Like many of us, I tend to keep company with people who largely share my worldview. So it's rare that my veganism has been the subject of true awkwardness or militant opposition. I certainly eat with a lot of carnivores, but usually it's, at worst, the boring hypocrisy conversation: what about this, what about that, those sorts of things. I have friends who want to have interesting, intellectual debates about what it means to make certain kinds of ethical choices. So in the way that Safran Foer says, I think my veganism is a way to actually open up all sorts of really interesting conversations. Even before I started thinking about consistency and identity—and sometimes I'm ashamed of myself for this—but I don't tend to be very political about my veganism at meals. I'm a vegan, and I eat vegan, and if you want to talk to me about it, I'm happy to talk about it, but I don't really want to talk about it. I don't want to talk about it in part because it's emotionally difficult for me. It's partly because I keep company with people I respect, and I hate disagreeing with them about something so fundamental. I would rather be blind to that sometimes, and to feel like I have more in common with my friends than maybe I do. I also don't want to have the same conversation at every meal. I understand that it's important, and that's why I feel ashamed of my reticence. I know that there are people who are better activists than I am, who are willing to have that same conversation every time they sit down to a meal; I don't always want to. But this also rather baldly comes back to what I was saying earlier about efficacy. It's not just that I get tired of talking about veganism; it's not just that it's emotionally difficult for me to have these debates. It's that I know that it doesn't work. I just appreciate the fact that when I eat vegan more people end up also having to, if they want to share a meal with me, or they try to make it easier on everybody and we go to a vegetarian restaurant, or whatever the case is, and I usually kind of leave it there. It's modest, but it's more honest to who I am.

SK: That raises the question about moving across different geographies, and navigating your veganism as you do so.

ND: Yes, it's in India where I have faced the militant, and not just tedious, responses. The nonsense about being in bed with the Hindu right because I'm not eating animals. To me, it's just so obvious that violence is not an appropriate response to violence. It just doesn't make sense to me. I see the worldview in which it does make sense, which is a worldview of hierarchy and normative value, but this brings us back to the "What do I know?" question. I'm happy for people—I'm, of course, being sarcastic—who know what the hierarchy of value is. But I feel I need to answer the question "What makes me so special that I perform my politics by taking another being's life?" At the end of the day, the answer is pure metaphysics. What makes me so special? Well, I am. Because I say so.

Being vegan instead of vegetarian is actually useful in India, though. The Hindu right is as appalled by the rejection of milk as they claim to be appalled by the slaughter of cows. So while vegetarianism is associated with right-wing politics, veganism is, at worst, associated with being Westernized. It has a different valence than vegetarianism.

SK: Could you talk us through your typical approach to buying or otherwise procuring food?

ND: Sure. I'm vegan, but I'm also just into being obnoxiously healthy. Those two things, they go together in some ways, but not necessarily in other ways. The veganism plays an important role in terms of setting rules; there are just things I don't buy because they're not vegan. I tend to do the classic Pollan thing of only buying from the outside aisles of the grocery store, and buying very few processed foods. So those are some rules. But I also had an important insight a few years ago that rules can be counterproductive to a healthy life. For instance, I had a rule for a while about not having carbs after lunch. But I was hugely unsatisfied—I love carbs—and so I ended up, ultimately, eating badly. If you're just blindly following rules, you're not actually living a life, you're not living in your body. So besides the vegan rules, the only other one is that I do not deprive myself of what I want and need. That's the only rule I found that I actually have to follow. Of course that also means having to be in touch with who you are as a physical being, and that's precisely one of the things that a rule-bound framework prevents. I don't need to be in touch with who I am—and we can extrapolate this to queerness, to veganism, and all sorts of things. I don't have to be in touch with who I am as a person, because I already know who I am. I'm a vegan and I like women. I often find myself maybe chucking those rules, at least hypothetically; and just paying attention to "What do I want, what do I crave?" ends up being the only rule to follow.

SK: This is a really good segue to the next question. You clearly try to avoid universal moral prescriptions in your work and are at pains to emphasize the importance of openness, of not knowing in advance what one will argue or what kind of politics one will pursue. This is easier said than done—how do you resist the urge to fall back into predictable or comfortable ways of thinking?

ND: Well, the first thing is that I think about how much I hate it when other people do that [laughs], and how impossible we become. Few things are more important to me in life than the value of a good conversation. You can't have a good conversation with someone who already thinks they know everything, and they know how everybody should live. I feel like all the beautiful things in the world—expression, discovering new ideas—it all emerges when you allow yourself to be dumb, to not know everything. Maybe I police myself a little bit; I try to catch myself falling into militancy or, worse, self-satisfaction. I try to catch myself. Interestingly, several of the animal activists whom I work with in India are into *vipassana* meditation. This is also related to the question you were asking me earlier about what it meant to be around so much animal suffering. Well, this is how they deal with it. They practice meditation. It started out for me as a way of understanding my interlocutors better, going on a ten-day *vipassana* retreat. But it became something much more than just learning how my interlocutors do things; it became a central practice for myself. That's where I started thinking about deprivation and excess. I also find that a daily meditation practice is, at the end, the most useful way to avoid moral prescriptions. But some might say that itself is a moral prescription!

SK: Do you see yourself continuing to pursue research on animals?

ND: I think I'm a serial monogamist in my writing [laughs]! I'll write on queer politics, and then run away as far as I can. Then I devote myself to thinking about animal politics, and then I'm sure I'll run away as far as I can. But that's okay. It's taxing writing about something you care about. And this goes back to the very first question you asked about motivation. I think a lot about David Graeber's essay on animals in play, and for writers or academics, this is our version of play. And play is interesting insofar as it's not about the repetition of the same, but about horizons, affect, the emergent. And of course, I don't know how many times I've mentioned queerness in this conversation, so obviously I haven't gone very far away from that. So it's interesting to see how ideas stay the same but perhaps in new and less recognizable shapes. My next project won't superficially be about "animals," but at the level of substance, I'm sure it will be infused with all of the things I'm learning and thinking about right now—maybe it will be animal rather than being about animals.

CHAPTER 5

Justice and Nonviolence

Maneesha Deckha

A law professor at the University of Victoria, Maneesha Deckha has always been interested in questions of justice and nonviolence from an intersectional, postcolonial perspective and credits feminist theory for catalyzing her thinking about human–animal relations. Deckha discusses her strategies for integrating animal studies into her legal pedagogy and why she sees teaching as a potential vehicle for social change. She also shares her experiences as a vegan parent of a vegan child and the deeply affective commiseration she feels when contemplating the routine separation of baby animals from their mothers for milk and meat production. We interviewed Deckha by Skype on October 28, 2016.

VICTORIA MILLIOUS: Can you tell us a little bit about yourself?

MANEESHA DECKHA: I moved to Canada when I was two. So that whole new-immigrant experience—my family is originally from India—was I would say very central to my formation. We lived in a new-immigrant enclave in Toronto, so I grew up in a public school system that was very multicultural and multiracial at least for that time—the '70s in Canada. And so I feel like I had a hybrid upbringing in terms of a hybrid identity, being South Asian, Canadian, Indian-Canadian. Also my cultural-religious background is Hindu, so we also had that aspect growing up. Hinduism is so diverse and allows all different kinds of manifestations of adherence, in fact it doesn't even require adherence, so not in those terms, not like going to a temple or *mandir*, but just the sense in the home, especially with my mother in terms of celebrations, like Diwali, which we're celebrating now, things like that. So that was very formative up until high school.

I encountered early experiences of overt racism, obviously there was covert racism that perhaps as a child didn't really appear to me consciously, but I definitely remember the overt experiences from growing up in Toronto. They were just shocking. But really I developed a racialized consciousness when I switched schools because we moved north of the city at that time to Richmond Hill, which now is very East Asian, but at that time in the late '80s it was very white still for a Toronto suburb, or not even a suburb, it was a lot of farm lands then, so really, really different than it is today. And so I went to a school there that was really different from the North York community I had come from in that the racial composition was really white; it was also really Anglo and really affluent too.

So you know, that really changed my whole identity. I would say I was really extroverted before, a real leader in all my classes and schools. I was co-valedictorian in my junior high and then I came to high school and I just quickly wanted to blend into the back of the wall and not be noticed and it took a few days to realize what was different about the school. At the same time, I always received academic accolades and that continued at the high school level, but I really tried everything to fit in with my peers. And so the first two years were like starting from scratch, building up networks and friendships. My last years of school were socially enjoyable. I didn't have one of those alienated high school experiences, but really my personality had changed and I was just not a leader except performing academically and on an individual basis.

And then I went to university, to McGill, for undergrad and I stayed out of university housing for a variety of reasons, so I was not in that student environment. And it was there, through social justice clubs, through courses in humanities, and through an arts undergraduate degree, that I came back to myself as I was as a child before that really white high school experience. This was 1991 to 1995, and the mood in social science and humanities in the critically oriented courses I was taking was very deconstructive and that's where I took on a feminist identity and also started learning a lot about postcolonial theory, queer theory, Marxist theory, all the range of critical social theories. And I was also away from my parents, though I was near my brother who was at that time a big influence on me. We lived together, but I was away from that high school environment and able to stake out an adult identity that was definitely more politicized and aware. Academically I always enjoyed school, and university even more so because it was self-directed. And I had early ideas that I wanted to be an academic; I wanted to be a professor. I would often pause in my undergrad and think, "What do I want to do with my life? What really motivates me? What can I imagine spending hours and hours at work doing?" And it was really the questions about social justice, especially around race and gender, that were empowering for me and so I started to think about what type of career I could have. At McGill at that time there wasn't a major in women's studies, and I worked with the Centre for Research on Teaching Women to petition the Dean of Arts for an ad hoc honors program that I could take, that I could create it. I had the full support of the chair of the center, but I remember that being denied by the Dean of Arts, the reason being that if the faculty was going to allow an ad hoc program, it was admitting that it doesn't have a complete program. And it didn't. And then a few years later they did start offering a major or honors in women's studies. But anyway, I went ahead through the existing flexibility of an interdisciplinary joint honors designation and ended up doing a joint honors in anthropology and political science, focusing on cultural anthropology and political theory with what was then available for women's studies, which was a minor.

VM: Was this also the time when you began thinking about human–animal relations, or did that come later?

MD: No, it was actually through feminist theory courses that I began thinking about that. There were a lot of critical theory courses and then, as we were questioning all these classical intellectual traditions, I thought, well, why don't we think about the human as a social construction? Nobody's really talking about that, extending that, thinking of species as a social construction and then thinking of the human–animal divide. And I was motivated to think about that because of what was going on personally in my life. I did not grow up being a vegetarian in my family. We didn't eat cows, but I grew up eating other things. And, like I said, at that time I was very close with my brother and he had gone away to university and had come back after his first year being a vegetarian and that really influenced me. To explain why, he showed me some animal rights pamphlets and I read about a mother–child narrative of a cow and her calf in transport and that was it. I just thought "Oh my god! I had no idea." I felt an amazing passion for what was going on so I became a lacto-ovo vegetarian then. That was in high school, and then by the time I went to university myself and was feeding myself out of residence and cooking, that's when I transitioned to eat vegan. And then in my classes I had these questions about well, what about the social construction of the human and the animal? And then again in courses where you had control over what your term paper was, or other less major assignments, I took on these questions, whether it was a literary course or political theory or anthropology. So I was able to nurture that as an undergrad even though there was no specific course that I was aware of on human–animal relations at all in social science humanities, or that looked critically at this as a major theme of the course.

So then, thinking about careers, I gravitated toward law. Like many students coming in, I really didn't know what I wanted to do with my life, but at least I knew a law degree would provide some type of legitimacy, credibility, and security for a future career. And so I set my sights on law school and I ended up going to the University of Toronto after my undergraduate degree. Again, that was a bit of a homogenizing experience because in the first year the curriculum was all set and there was very little critical theory, at least as it was taught by the professors that I had at the time. And the University of Toronto Law School was very much a high school environment because all your needs and courses were in one building, and you moved through all your classes with the same 80 students. So I felt I had been kind of catapulted from the critical theory, activist echelons at McGill, back to a very affluent, mostly white community at the University of Toronto, where it was a very liberal humanist paradigm in the classroom. So in the first year, being committed to it was a struggle. I was committed to my degree and education, but really *liking* the courses was a struggle, and fitting in was a struggle. I also lived at home then with my parents, so I wasn't able to immerse myself socially as much as if I had lived with others. And the nature of law school is that, even though the second year is officially elective, you feel that some courses are mandatory, because you can't make sense of all the core courses you have without taking certain others. So again I was taking these courses I felt I needed to take to be a good lawyer, to pass the bar, to get articling positions. It was really in the third year that I finally felt I had room in my schedule for the seminar theoretical courses

I wanted to take. And when I did that, I came back to thinking about being an academic. I must also say, the first year of law school is a very sobering experience. Almost all of the students at the University of Toronto are used to straight-A averages and high achievement—and all of a sudden you are graded on a curve, and only 10 percent of the class is scoring above B+. So it totally deflated my confidence about being a law professor and I didn't think that I was going to go into legal academia. But when I came into the third year and started taking the courses I really loved, then I saw my grades go back to the level I was used to. It was also a smaller environment and there were seminar, writing-based courses rather than the 100 percent exams which dominate first and second year. You get to know your professors, and I had very encouraging professors. I have to credit Jennifer Nedelsky, a very prominent relational feminist theorist then at the University of Toronto, for giving me the confidence to think about grad school because she basically told me to do it. I confessed my fears about my transcript to her and she put my grades into context, that these seminar-based courses were much closer to what grad school requires than the exam-based courses.

It was actually at the University of Toronto Law School where I did a directed reading about animals as property in the law with Professor Craig Scott. I also had a variety of courses where I was able to explore feminist issues and feminist animal issues, so I really enjoyed that time. And then I had this law degree and the whole mass thinking from the students is, "Of course you're going to article, of course you're going to practice," so I felt caught up in that. I also wanted to practice for a little while and see what that was like, and get called to the bar. So I did that, but I kept in contact with my professors, who became my referees for graduate school. I got called to the bar in 2000, worked as a lawyer until August 2001, and then went to graduate school. I did my LLM at Columbia between 2001 and 2002. There I took a lot of theoretical courses. I also did a directed reading program called an LLM thesis with two prominent feminist theorists: Katherine Franke, who is at Columbia, and Reva Siegal from Yale University Law School, who was visiting that year. I didn't do an animal project because I felt, based on their interests, it wouldn't be their first choice of what to supervise. So I did a project looking at the cultural defense in law, about a series of papers that constituted a literature by then, that had come out among feminists about how courts should deal with cultural claims, largely (but not purely) by males accused, in cases involving domestic violence. The perpetrators argued that their culture should be taken into account when they are being assessed for whether or not they're guilty of something. It's usually seen as an excuse for what they've done. I looked at some of the feminist writings on that, which I saw as very divergent approaches, to assess them and suggest a path forward. So that was my LLM thesis.

Again, at that time I was suffering—as many women do—from lack of self-confidence as to actual abilities. I didn't think I was in a position to start applying for academic positions. I felt like I didn't have any real publications; I just had one student publication; I was still in an LLM program, maybe I should be in a doctorate . . . a host of reasons were in my head. But then I saw these other Canadian students applying to Canadian positions, and I began comparing myself with them. One thing the University of Toronto Law School does subtly impart, for good or for bad, is that it is "the best law school in Canada and any other law school doesn't compare." So I saw students from other law schools applying and I was like, "Well, maybe I should, because I did go to the University of Toronto. . . ." That gave me some motivation and confidence to go, "Okay, let me just apply and see what happens." It's a good thing I did, because I ended up with two interviews and very fortunately with two job offers—both of which were based on job talks about animal rights. Because with the interview process, I basically got these interview offers and then I had to travel from New York for an interview in Victoria, and then an interview in Ottawa within two weeks. And that was just totally traumatizing because I was so scared of doing a job talk in front of faculty members, and I also knew going back to Canada to talk about American law, which is what I was immersed in at that time, was not going to be popular. So I had to think, What could I create a job talk on that I knew like the back of my hand, and that would not be rehearsed or heard to death by these scholars? So I settled on animal rights—against some advice that it was actually a death knell to do this. I talked about animals, and a critical reading of intersectionality as a theory, and why it didn't apply to animals at that point; I talked about species difference as something that should be factored into an intersectional approach. That's how I got my job offers and how my academic life started.

VM: What do you see as the primary purpose of your work, and why do you do what you do?

MD: To make a social impact; to make a social difference; to actually bring about some type of change. I'm really motivated by my personal sense of what compassion and justice entail, and that has always influenced my intellectual interests, even before they fully enveloped the question of animals. As an academic, I realize that most people reading my writings are just other academics; the audience is very limited. But I try to think about teaching to students as well, those who aren't ever going to read my research publications, through my core courses in the law school. I feel that part of my work is to contribute to a field that is trying to make social change and trying to challenge intellectual inquiry. And in the scope of those inquiries to affect other academics, and academia as well. But also to reach students who are presumably going forward with legal careers, who are going to have some type of position of influence, as legally trained people usually do compared to the rest of the population. I'm hoping to influence them as well to think critically about human–animal relations and bring that into their lives. I really became an adherent of intersectional theory, and when I realized it was so human focused I really wanted to think critically about why that was as it seemed so obvious to me that its tenets should critique anthropocentrism and make visible the connections of human-based oppressions with species hierarchies.

VM: I like what you mentioned about teaching. Could you describe how your work resonates with your students, or does it? What experiences have you had teaching your material, and what kind of reception do you typically get?

MD: When I first started it was in a core first-year course, which typically had a small cohort that was outwardly hostile to critical theory and critical thinking. I used to teach a famous American article in critical race theory when I taught property law. So property law is a mandatory first-year course which I taught for my first six years, and for many of those years I taught this article called *Whiteness as Property* by Cheryl Harris—an article that tries to explain why we can understand the racial identity of being white, at least in the United States, by focusing on property. And one of the first things you teach in property in the common law tradition is that property is a *right*, not a thing. We tend to think of property as "things," but we what we learn in law school is that property is a right. And it's intangible; it's a conceptual way of thinking of things. This is what the article about whiteness is talking about. And that class always incited a lot of student resistance. I'm laughing about it now, but teaching it as a young junior professor was not fun. I did it really out of a commitment to critical race tenets, a feeling that it should be taught. And I never thought there was going to be something that would enrage students more, but when I started talking about animals, I was like, "Oh! Yes, here's a topic that is more difficult to teach."

I have to say at that time, and also today, the University of Victoria Law School is quite a white community. It's quite a progressive school, but it's also quite a white community. In my courses I used to do modules toward the end of the year about thinking really critically about what it is we say is property. Because in the common law, the law basically divides us into two things: You're either a person or you're property. So I would do modules inspired by feminist theory about, What do we say absolutely falls into the category of personhood and can't be commodified? This was inspired by theories about commodification of eggs and sperm in the human body, and whether women should be able to sell certain things in their bodies. Following that module I would then flip the question to say, Well, what is it that we normalize as property? What do we always already think of as property but maybe should think of as being in the personhood realm? And then I would introduce the topic about nonhuman animals as persons. The first time I did this, you could hear laughs by some of the students in the classroom, like how could this possibly . . . ? And to use the term *nonhuman animal* at that time, not animals, elicited all kinds of guffaws and laughs and snickering. So then I had to come back into class and be quite mainstream about it and just arm myself with these big names from big American law schools who talk about these issues, and bring in the books and talk about their theories. But it was very dispiriting to do year in, year out—so I didn't do it year in and year out. Or I would enlist the support of respected white male colleagues also teaching about property in other sections, to have the same syllabi and for them to include some material on animals. That was a strategy about how to get things across. It's a continual strategy, not just on animal issues but at law schools in general, I would say, about how to introduce critical thinking in the first year without eliciting backlash from students who just want to be lawyers and tell you that if they wanted to critically think they would have gone to grad school and not law school. It's a common refrain you hear from students. And it's a total pressure-cooker situation in the first year with hundred percent finals and so much riding on transcripts, that the anxiety spreads when you hear grumblings from a small cohort about what's being taught and what's not being taught. Yeah, it's usually hostile. But then, my first seminar I started in the law school was called *Animals, Culture, and the Law*. And with students in seminars, who are self-motivated to be there, it's always been a very positive, great teaching experience. And really one of the highlights of teaching for me.

VM: Sammi and I are sharing looks while you're speaking because as sociocultural theorists and teachers in a kinesiology department we have a somewhat similar challenge in that many of the students come here because they want to practice in medicine or sports therapy. So trying to introduce the critical theory and the sociology courses to persons who really don't have an interest in that and don't want to cultivate an interest in that can be a challenge.

SAMANTHA KING: Is there a key concept that guides your thinking about animals and your other areas of interest?

MD: I would say, in terms of interests and personal motivation, the whole idea of justice has always motivated me—to go to law school, and to think about making law reform and change. In terms of intellectual analysis and writing, I would say it's probably the concept of nonviolence—what is violence, what does it look like, what would nonviolent, peaceful interspecies relations look like. This whole idea of nonviolence informs all of my interest in theory, and how to live nonviolently and be free from violence.

SK: Can you discuss what you see as the potential for postcolonial, posthumanist feminist theory in approaching questions related to nonhuman animals? We're asking here, in part, about what you see as the particular contribution of your work and the lens that you bring to it—both within law and outside, since your work is read far beyond the legal field.

MD: Thank you. The strand in eco-feminist theory that is variously called "vegetarian eco-feminist theory" or the "feminist animal care tradition," and which is epitomized by scholars like Carol Adams, Josephine Donovan, Marti Kheel, Greta Gaard, and Lori Gruen, for example, was very much an intellectual home for me when I first started thinking about feminist theory and animals, because here you have feminists talking about individual animals and why they matter. As much as I align myself with that theory—and obviously the intersectional critique of it—I always felt that contrary to anti-essentialist critiques that I also worked rigorously with in the 1990s and early 2000s, that this strand of eco-feminist theory was more rooted in the cultural feminist/radical feminist realm. So while these feminists were talking about racial oppression and class issues, these issues weren't central to the analysis, as gender was. And through anti-essentialist critiques my worldview has always been that gender is always already informed by other vectors of difference.

In my undergrad years, I was very much informed by postcolonial theory—the whole critique of Western frames of viewing everything. And when I became vegetarian, that wasn't an issue for my nonvegetarian parents, perhaps because of my upbringing in a more culturally Indian Hindu tradition where vegetarianism is more the norm. There was more of an issue with veganism because it is much less familiar, but vegetarianism was so normalized. So I always felt that these Western traditions about how to view animals in the Judeo-Christian ethic, that was a key problem, especially in Western societies, as to how animals are viewed. So I felt that a postcolonial reading of human–animal relations was central to a critique of animals. And I didn't see that as centralized in the feminist care tradition. So in my writings I am influenced by the strand of eco-feminist theory, but I also very much feel that postcolonial analysis is necessary to properly contextualize issues; to understand how food issues, food politics, and other issues involving animals are not only gendered, as feminists in this tradition talk about, but also very racialized issues. You could probably have an intersectional analysis of what's going on in terms of how humans treat animals and encode animals and constitute their own identities through animals and animality. I credit postcolonial theory with really bringing the concept of "otherness" into the Western intellectual sphere, so I invoke it for that reason as well. Through postcolonial study I am also mindful of the debates about not having universal positions, or universal positions being seen as anathema to postcolonial theory, looking at things in localized ways. Then that brings up the question, Well, what about marginalized cultures, human cultures, indigenous cultures that use animals for food, and how to think of that. I feel a postcolonial reading brings those issues more centrally to debates than readings which don't bring in critiques of Western ways of knowing and seeing.

SK: We're going to move now to the more food-oriented questions. You've talked a little bit about your transition to veganism, but maybe you could discuss your current relationship to consuming animals and to what extent has this been an issue or preoccupation to you. Have your thoughts or feelings changed over time?

MD: Yes, I am still a vegan. I can't really say my thoughts have changed over time about that. It's not really a preoccupation for me; I guess these days I'm more preoccupied with the judgment I receive for having my young child be vegan. I've become aware of the sensationalist media stories about vegan mothers being charged with neglecting care and things like that. With the baby, I have to say that I don't go around announcing that he is a vegan. For myself, I'll go to a medical practitioner and say I'm a vegan. But I won't say that about the child just because I feel the biomedical profession is largely against my beliefs about parenting. There are so many areas of clashing that I don't generally offer information, unless I feel it is relevant, about how I parent at home. I understand the pressure that mothers are under, in terms of getting scrutinized. I think now in the last five years there is more mainstream awareness about veganism. I wouldn't say it's *mainstream* at all in terms of support, even by nonvegans or vegetarians, but more people certainly know what it is. But maybe it's just this heightened sense of protection toward my child. I do worry there could be some ridiculous claim brought about raising humans vegan that gets morphed into some huge legal issue for us. I think that's what makes me really nervous about talking openly to authoritative figures in various professions about his veganism. To my friends, obviously from my principles they know what I am about, and when we socialize, the child is eating what I am eating. And as they do over dinner tables, these questions about vegetarianism and veganism come up endlessly, so it's more known. But I don't go to my GP and talk about my child being vegan.

SK: While we're on dinner tables, I wonder if you could talk about your approach to buying or otherwise procuring food, and how that has changed or not, now that you're a parent.

MD: Sure, yes. I try to be conscious of all my choices. When I moved to Victoria for my job in 2002 I quickly saw it was an organic capital of Canada. At that time I didn't feel like the price disparity between conventional food and organic food was as significant as it was when I was a student in Toronto—by then I was earning. So I started eating organic; I don't exclusively eat organic—I'm still price-conscious to a certain extent—but we do eat a lot of organic food. And in British Columbia I tried to buy local, from farmers' markets in the summer. All those issues: trying to eat less processed, making things at home—mostly to control other factors like salt and so on. I would like to get into the realm of growing my own vegetables; I haven't yet, but it's an aspiration. But yes, we try to eat locally, organically vegan choices. And when it's not prohibitively expensive, to buy certain commodities fair trade if they only come from international sources.

SK: You've said that your perspective and practice of veganism has been pretty constant for a long time now; is there anything you'd like to add about the relationship between your writing and eating practices?

MD: When I became a mother, and I started breastfeeding, I started becoming more acquainted with the type of parenting I am following, which I guess would fall under the umbrella of attachment or natural parenting. At this time I became really aware of these issues: the importance of having your infant close to you, what they need in their first year in terms of responsiveness and connection and stimulation, this experience of affect attunement. I became vegan originally because I learned more about the dairy milk industry, because I realized a connection to the veal industry, but it was really because of this idea of drinking another mammal's milk that resonated with me. Like, why would we do that? No other animals do that. And then seeing how dairy cows were treated. But it wasn't until I became a mother that I really emotionally connected to the idea of: This is a mother whose baby is being taken away from her; this is a baby being ripped from the mother and being denied what is rightfully that baby's, that's their milk not a human child's milk or a human adult's milk. That is what the calf needs to live and she also needs to be with her mother to thrive. Even these words, I feel, don't give justice to this sense of this deep compassion I feel for this infant child bond now that I'm in it, and I just can't even look at the pictures of mother cows or pigs, such as pigs caught in gestation crates and away from their piglets. I think I posted one of those images on my Facebook account and commented, "If I wasn't a vegetarian before becoming a mother I certainly would be now." And it shows that separating moment and the trap moment of the mother that can't get to her piglet, or the piglet that can't get to his mom. I just find it devastating to think about what goes on for those animals now. It's endless misery, pain, and violence that all stems from the exploitation of female mammals' reproductive capacities. The emotional trauma for the mothers separated from their calves really became heightened for me, like my commitment to veganism—to not drinking dairy especially—has been fortified by this experience. That being said, it's still in the day-to-day moments of eating with others at celebrations or worrying that my child should have enough solids when we're out on the road and looking at the packages, like "Does this have whey in it, does it have this in it." I wouldn't absolutely rule out having something with a milk product in it, not something like milk or ice cream or cheese or anything like that, but if I felt really under pressure and that was what was available I might give him that. But then I also see that as a failing of my own commitment that I should really not, and I should have been better prepared by bringing something for the baby. I also feel tremendously guilty when that occurs since I immediately recall the images of mother-and-calf separation. I am also indelibly shaped in my compassion/empathy/guilt by Greta Gaard's closing narrative in her 2013 article in the special issue of *American Quarterly* focusing on sex, race, and species where she seeks to develop a postcolonial feminist milk studies and tells the story of the mother cow who has twins and hides one away from the farm to try to save that baby. She knows from experience that the farmer will take away her calf and proceed to milk her and so she hides only one calf, knowing the farmer will take away the other baby, who is indeed, stolen from her. I will leave it to your readers to find out the heartbreaking end to the mother's effort to save at least one of her calves. When I read that account I couldn't stop crying. It is an image of violence, pain, and trauma that I can't get out of my head even now and also really don't want to completely because I feel it holds me accountable to do something for all the animals treated so heinously, who are ripped from their kin and denied their most desired loving and social relations. So, yes, I really do feel more intensified with my commitment to veganism on a very personal level. Now if I try to talk about veganism to a mother—if they brought it up; I don't typically bring up these conversations socially myself—I would emphasize this relation, the mother–child issue inherent to the cow–calf relationship, in a way that I didn't before.

SK: One of our other interviewees referred to his veganism as a limited tactic. He was trying to draw out the differences between veganism as a political tool versus a personal ethic. I'm wondering how you view your veganism in relation to your commitment to social justice and how you would think about it in terms of ethics or politics.

MD: I guess I would say it's both. It's important to my sense of identity, to who I am as a person even if I weren't in academia or weren't engaged in what I see as politics. But it's also a very politicized identity because it's such a minority position; you're inviting at least challenge—if not ridicule and hostility—in social spheres that are not exclusively your friends. Or even among friendly people it can be a source of challenge in a way that other eating practices aren't. I do want to do it for political reasons; I would be vegan even if the rest of the world was vegan. So I don't only do it as a tactic to emulate change; it's important to me in terms of my ideas of justice and violence, and what my ideas of animals are.

VM: Can you tell us about a time when your dietary practices have been the subject of awkwardness, celebration, or hostility? Do you have a particular story you wouldn't mind sharing?

MD: I can't say that there's one that stands out for me, though there is a kind of redemption story. I was on exchange in California in my third year of undergrad, at San Francisco State University. I quickly found a group of friends and developed feelings for a boy, and when he heard about my veganism he totally ridiculed it in front of our other friends, basically calling me a cow—I didn't mind that association, but I knew he meant it in a disparaging way—because "all I ate was grass," and then we were still friends, and I still had a crush on him at that point. But then I saw him fourteen years later in 2008. We were both in New York, so we met up. He took one look at me and said something along the lines of, "Wow, you haven't aged a bit—I guess veganism is the way to go!" That was the first thing he remembered about me and I was like, "Oh, if it takes fifteen years to change someone's mind, it takes fifteen years." I mean it's common; that's probably the most negative thing that anyone has ever said to me, that was an insult on a personal level in front of me. Obviously through the years there have been awkward moments at dining tables about, "Why are you vegan?"—things like that. But nothing I can recall was hostile. A moment of celebration was my vegan wedding, where one of my friends who spoke at the speeches really talked about that. And as much as I engaged in the whole wedding industry, I tried to make choices around food and other issues about consumption that were more in line with my politics. That was nice to hear because I'm sure there were guests who were wondering what they were going to eat at a vegan wedding. So it was nice to get that public moment to acknowledge me for it.

VM: Can you describe a key dilemma or question that haunts you?

MD: I don't think it haunts me . . . but I find it odd and dispiriting that veganism is challenged as being racist or colonial. Maybe it's my upbringing in the Indian South Asian tradition of nonviolence; hearing about Gandhi, learning about Ahimsa at an early age even though we weren't vegetarian as a family, learning that in India the restaurants are called "veg" or "nonveg," which shows you the framing of what is normative and what isn't there. Vegetarian or nonvegetarian—it was just normalized; it's kind of an ethnic identity to be vegetarian. So to hear critical intellectuals frame it as an exclusionary choice, I don't think those critiques are well sustained when you take a global perspective and when you think about how the world eats and how different cultures eat. I understand the reasons for those arguments; I'm sympathetic to some of them, but it's never resonated for me because I've always felt that vegetarianism was an Indian tradition or at least an Indian–Hindu tradition and the whole idea of nonviolence is really something I've learned about for a long time. While I admit veganism can be an exclusionary choice in certain contexts, I think that's true of all practices. So always discussing and arguing about that critique, I find frustrating. Especially if you think about the levels of violence animals are facing, the immense difficulty of mobilizing humans to see such violence as violence, and what a minoritized position practicing veganism is, I feel that getting mired in that question is spending precious time on an issue that should be directed elsewhere. At the same time, I'm mindful of the critiques of feminist theorists who are not embracing anti-essentialist theory in the 1990s, saying, "There's so much feminism has to do; why are we focusing on these debates within feminist theory about whether or not mainstream Western feminist theory is essentialist." Of course I focused on those debates and I think it was important to do so. So it's an important debate to have, but ultimately I feel it's a dispiriting one. And my postcolonial teachings really motivate me to always leave my views open to critique and alteration; and the idea that universal propositions should be suspect, so I really try to grapple with that. So I'm trying to live by a nonviolent ethic, which for me means being vegan. I would like that to be promoted among everyone else, but then I have to think about whether that is a problem, given that not all people see violence in the way I do. And to constantly think about how I accommodate other positions, or grapple with them, or even offer a gesture of accommodation. Though I don't like that line because it seems like a recognition of a hierarchical relationship. Basically, how to have a nonviolent ethic for all beings, people, animals, other nonhumans and still allow for exclusions or exceptions is an issue that I struggle with.

CHAPTER 6

Doing What You Can

Kari Weil

A scholar of comparative literature, Kari Weil began writing about animals while researching the transformation of equestrianism and human–horse relations in nineteenth-century France. With an eye to feminism, sexuality, and gender, Weil's interest in horses led to new questions about animals and human–animal relations. Through both her research and teaching, Weil questions the strict boundaries that distinguish "human" from "animal." She discusses the role of language and literature in helping us imagine human–animal relations differently, and what animals might tell us if only we could understand them. We interviewed Weil by Skype on December 16, 2015.

SAMANTHA KING: Since we're interested in the connections between personal and academic interests, can you describe where and how you were born and raised, your formative experiences, and how you became an academic?

KARI WEIL: I was born in Chicago in 1954. I was raised in a family that had a strong gendered division between my father and mother. My father was very much an intellectual; he spoke many languages, had a huge library, and would have spent all day reading books if he could have—but he also had to run a business that he had inherited from his father. I suppose my love of literature, and my interest in literature and languages, came from him. My mother, on the contrary, loved the outdoors. If he was brain, she was body—this was sort of the running joke in my family. She loved to exercise and be outside. She didn't care that much about books, although she admired my father's ability to quote. This made for a gendered mind/body split that maybe was formative in some way; I hadn't really thought about it. I was very much encouraged by both of them to travel and learn languages, though reading came from my father.

In terms of how I became an academic, I had the luxury of my father's encouragement to pursue graduate study without him pestering me about "What are you going to become?" I know that was a privilege, and I'm grateful for that. He wanted me to do, in a sense, what he didn't get to do. I went to graduate school, not necessarily knowing I wanted to be an academic, but just because I loved thinking about ideas, thinking with books, thinking with words, playing with words. I kept telling myself throughout graduate school, "Well, at some point I'm going to quit, I'm not sure where this is going, I'm not sure if I'm smart enough." Lo and behold, I just kept going through and then got a job, and there I was, an academic. It wasn't something I had necessarily planned. I was a French major as an undergraduate, and then I did comparative literature. And that's probably very much influenced by the fact that both of my parents spoke French. We had French friends; I heard it a lot. My father was from a German background, and my mother was Norwegian, so I had a very European formation, which is also why I learned French and German as well, which they both spoke. And I found I loved French in school. It wasn't really anything more than that—I loved it and I was good at it! With no idea about animal studies—certainly at that point—my interests in graduate school were much more geared toward questions of gender and questions of language. I think that was the first interest, and that moved me into animals only later on, but certainly both of those remain prominent in what I do in animal studies as well.

SK: How did you come to write about animals?

KW: My first book and my dissertation focused on questions of androgyny. I was looking at literary depictions of persons who appeared to transcend sex or transcend gender—although *trans* was not yet a colloquial term—and how that played out in certain novels. Specifically,

I was looking at these issues in nineteenth-century France and Germany, and eventually twentieth-century England. It was actually part of my research for that book that I came across women in novels who cross-dressed in order to ride horses. I have loved horses since I was a kid when I went to horse camp, and I have tried to keep riding ever since. When I was coming up for tenure, I told myself if I get tenure I'm getting a horse, because I'll probably never have a family, and my horse will be my family. I got tenure, got a horse, and then I got married and had a kid and moved to California; I had been teaching in North Carolina. When I finished the book on androgyny, I decided I could study something I loved and so I began looking at images of women and horses, and the figure of the Amazon or woman rider in the nineteenth century. The research was still very much related to questions of feminism, sexuality, and gender, but now all woven into relations with horses and the ways that women riders were perceived.

My research took me to examine how relations between men, women, and horses changed in the nineteenth century, especially as riding changed from a military and aristocratic privilege to a bourgeois sport. I also began researching one American woman who was a stage actress, first in the States, and then in England and in France. She became the hit of the theater, by riding a horse almost naked on stage. She was literally the first "Miss Mazeppa," the bawdy performer later immortalized in the 1959 Broadway musical *Gypsy*: "Once I was a schlepper, now I'm Miss Mazeppa." Eighteen-sixty-six seemed to be the "year of the horse," with everybody talking about Menken and her horse act. She was also a Jew, and I noticed how talk of her and her erotic ride blended into contemporary discourses of race and breeding—all these questions of sex and gender and race started coming together in odd ways. Then I also found that the year she was the hit of Paris was the same year that they legalized horse meat for human consumption. Suddenly this project that was just on women and horses really became a project about changes in human–horse relations and the changing status and meaning of the horse at the time. How was it, I asked myself, that one could fathom the idea that this horse, who was increasingly becoming a pet for many French—if also a beast of burden for others—was also considered edible for humans.

While I was working on this horse project—which I am now close to finishing—I changed my job and moved from a French department to an arts school that didn't have a French department and became chair of what was their critical studies program. One of the required courses was called a "methods" course, that looked at how we learn through the disciplines—not unlike what my department does now at Wesleyan. I had been reading a lot beyond horses, in what was *starting* to look like a field of animal studies. It really didn't exist when I had started the work on horses—I had never heard of it. And I began to get really interested in other questions about animals more generally, and human–animal relations, and so I started teaching a course called Animal Subjects. Since these were art students, I thought I would grab them with something that might interest them—either because they'd grown up with animals, or worked with animals. And actually it worked wonderfully! Most of the students who took the course were fascinated with it, and I could do all kinds of things that I couldn't do in a French department, including taking them to the zoo. I met some wonderfully interesting and engaged scholars who were docents of the zoo and taught us about zoo habitat, zoo history, and also how they really wish zoos didn't exist. But if zoos did exist, this was where they were going to work so they could make things best for the animals, etc., etc. So the class was very interdisciplinary and the students did all sorts of interesting projects. One designed aquarium habitats because he had a lifetime interest in fish and wanted to construct what could be a better aquarium than those being sold. One student from Hawaii wanted to investigate totems in his family and his culture, and the human–animal relationship in a totem. I had a lot of fun with this course, and I learned a lot. I think that the students really learned a lot, too, and it helped them think of their own work—sometimes in new ways.

It was in this course that I started teaching Coetzee and have been ever since. I taught his *Lives of Animals* first, and then I started teaching *Disgrace*. When I read *Disgrace* I was just *dumbfounded*. I was dumbfounded by Lurie's person; I was perplexed, and fascinated, and intrigued, and puzzled, by the ways that racism, sexism, and speciesism were intersecting, but not always mutually supportive. I was trying to figure out where Coetzee wanted to go with these different "isms," and I didn't understand how Lurie could put down this dog whom he loved, why that would be ethical, and why it could be in some regards. It was reading *Disgrace* that really started me on a new project. I had to write on it because I was so puzzled by it, and I wrote a number of papers that really got me going toward my book *Thinking Animals*. And then I began organizing my classes a little differently, around different questions, questions which were historical, but they were also ethical, and they brought not only gender, but also sexism, racism, and speciesism, together. I tried to get my students to think about these intersections.

SK: Could you talk about what you see as the particular contribution of literature to the study of animals and multispecies living?

KW: It's funny because I just had to write a brief piece about this. I've been working with a woman in California who is creating a museum of animal history. Along with a couple of others, I have been working on the literature section, and that question came up, "What can literature do?" For the short blurb I wrote for that, I thought immediately of Derrida's statement which goes something like: "Thinking concerning the animal, if there is such a thing, derives from poetry. It is what philosophy has, essentially, had to deprive itself of." Here is a thesis. And I think poetry could refer to literature, more generally, and that what Derrida is suggesting is that literature can imagine and represent human–animal relations, and more specifically, animal thinking—not just our thinking about animals, but the thinking that animals might do. Literature can do that without being fettered by the kinds of truth claims that philosophers or scientists have to abide by.

I think that literature can imagine things without having to claim that they're true, and I think it can also *imagine*, realizing that they may not be true, and that's important as well. It is both imagining what might be true, and at the same time taking stock of the limits of our own imagination, and how it may be shaped by our own human language. Some of this connects with my earliest work on feminist theory, and French feminism, because the question would often come up, "Is language patriarchal?" Not just, "Is it specifically human?" but is it patriarchal? Can language express women's desires? And what must we do to language to bend it, or shape it, or rethink it so it can fit a different body, a different desire, a different makeup? I think those kinds of questions are specifically literary questions, and that writers often think in those terms of, how can I both represent an "other," and remind my reader that this is an other—and maybe not one who can be easily understood, or even should be understood. How do I get my reader close to knowing there is an other there and, at the same time, realize that we need a skepticism about how much we know about that other, and a skepticism about how much we should even try and invade that other's being, and way of life, and way of thinking? Wittgenstein didn't claim that animals have nothing to say—just that we might not understand them. Literature, too, reminds us that they probably have a lot to say, even though we may not want to hear it, and we may not understand it.

ISABEL MACQUARRIE: I'm wondering if you can describe your relationship to earlier feminist work on animals.

KW: The question makes me think of somebody like Carol Adams, whose work I have used and found really helpful. Much of her work is groundbreaking, especially her notion of the "absent referent," and the ways in which our language, first of all, but also our practices, can disguise what it is we're talking about when we talk about meat. This is obvious in relation to questions of eating, how we call it "beef" rather than cow, or "pork" rather than pig. Even our packaging in the grocery store contributes to the disguise: What's wrapped in cellophane doesn't look like an animal. I liked Adams's book on the pornography of meat, especially for the images. It spoke to a part of my feminism, and I found it smart and inspiring. I had students read some of her work when I was in the arts school, and I still teach some—especially her pieces on animal experimentation. But when I taught her work on pornography, for example, I had negative reactions from my students, who either found her feminism too strident, or thought that she blamed men for our current situation with factory farms. I did find that some of her work seemed to reinforce a binary between sexes that made it difficult to deconstruct or question the binary between species. Much of my work, in contrast, was about undoing strict binary oppositions, undoing strict boundaries between species, and questioning the ways in which those hierarches and boundaries had been constructed.

That may be why the ways in which my feminism and my animal work come together were more strongly influenced by French feminists, even if they did not specifically talk about human–animal relations, although they have since then. I am referring to thinkers like Luce Irigaray, and certainly Hélène Cixous, who talks a lot about her relations with her dog and her cats. What I found in these feminists was an attention to the psychological reasons for the ways we "other" animals and humans alike. Concepts like abjection that come up in Julia Kristeva's work were similar to anthropological terms discussed by early feminists like Mary Douglas. Both describe the ways in which we construct boundaries between clean and unclean, between self and other, between self and the world, and the psychological, rather than existential, mechanisms that support those processes of boundary making. There's less about our relations to specific animals, or to food, for instance, in

French feminist work. But I think it's been more influential in terms of how we think of ourselves, or don't want to think of ourselves, as animal, and the ways in which our own animality is, and is not, represented to ourselves. That is what I find interesting in what French feminists have done. Carol Adams and Josephine Donovan have done groundbreaking work, but I came to them later than I did French feminism. There's something about French feminism that was always more formative for me.

IM: You've alluded to a few concepts that have influenced your work. Is there a key concept or idea that guides your thinking about humans and animals right now?

KW: I have a few. I'm somebody who rides, and as an adult I've always had a dog. I didn't grow up with animals—my mother didn't want them—but I try to have as many animals as I can around now. I often think of something Vicki Hearne writes: She basically says we have to earn the right to say, "Fetch." I've used this idea to talk about my teaching philosophy too. You don't just *tell* students what to do; you have to earn the right to teach and earn their respect, so they want to do what you ask them to do. I'm not sure I'm always successful, but I think it's a good concept and a good idea to live by: that we don't have a natural right to make anybody do something for us—we have to earn that right.

The other idea in my work that I keep coming back to is something that Derrida wrote. I've found his writing on animals to be some of the most provocative; it's not always easy, but it's made me think a lot. His simple statement about "the animal"—it "is a word, an appellation that men have instituted, a name they have given themselves the right and the authority to give to the living other." It's interesting—this idea of "right," of having a right to give a word. Whether we call a dog an animal, or a human an animal, we use this term. And it could be *animal*, it could be *dog*, it could be calling somebody a "bitch." The right to call somebody something is an idea that makes us question our categories and their origins: What are the reasons we have lumped certain animals or certain peoples together into one category? There may be obvious reasons—simply physical similarity—but I think there are also reasons beyond that, and that's what Derrida's phrase gets at. Because with the right to call something an animal, as he continues, comes the right to kill them, eat them, murder them—well, to kill them is to murder them without having it called murder. Because we call them *animal*, we can kill. It's not unlike other terms that have been used—whether it's *barbarian* or *enemy*—to justify killing other humans. I think that's, again, this issue of language that I've always been interested in. Why lump these creatures together? For what purpose are we doing that? Because it's our language that's doing it, but we do it for a reason.

I think the third concept is maybe more related to the interests in *Messy Eating*. I did a virtual classroom with Jonathan Safran Foer in a class I taught here a couple of years ago. Thinking Animals, I think, was the name of the class. We did a virtual classroom with people all over the world; it was an interesting exercise. I had students in my class whose politics and ethics were quite diverse, but when Safran Foer and questions of eating and so on came up—you know, do you have to be vegetarian, do you have to be vegan, do you have to be this in order to be ethical?—I think they all breathed a sigh of relief when he said, "It's not about an identity; it's about doing what you can do." I think about that a lot because, for me, too, I want to believe it's not about my identity, or the purity of my identity as being something, rather than trying to do as much as I think I should do, and can do within my circumstances, and that others can do within their circumstances, to better the lives of both nonhuman *and* human animals.

IM: Could you tell us about your current relationship to consuming meat or other animal products, and how this has evolved or changed over the course of your career?

KW: So probably starting with my work on eating horses, I stopped eating meat. I had not been a vegetarian before that, but I couldn't eat meat anymore once I started thinking about it. So it's certainly been my work that has influenced my habits, more than the other way around. I find that's important because my students often resist being told what to do. I feel that, perhaps like me, they need to come to their ethical decisions on their own, or they may resist and turn against them. It was while I was working on the horses and eventually teaching Coetzee that I started reading more and more about factory farming and paying attention to the evils in industrial agriculture more generally. I had a debate in my classes a few years ago, where I put meat eaters on one side and non–meat eaters on the other and had them go at it. As a professor, and as a teacher, I don't want to tell them what to do. I want them to come to think about it and hopefully come to some understanding on their own terms of what they think they should do, and whether they should do something about the plight of animals in our world.

That said, I suppose I would call myself—if I had to—I'm a "kind of" vegetarian who sometimes eats seafood. I'm a messy eater, I think, because I'm not *purely* anything, but I'm conscious of what I do. I think the messiness comes from the fact that my friends, my family, are not vegetarians. I think Elizabeth Costello also saw that there was something to love in these people, and that, "They're not evil!"—but she can't help thinking of them as evil because they eat meat. And I don't want eating practices to be the defining element in my relationships with others, because I think that would also be harmful. I don't want to be the person who spoils the party by saying, "Oh, I can't eat any of that." At the same time, I do think it's important to let people know that there are other ways of being. When I can, I will resist; I've rarely had a problem saying, "Sorry, I don't eat meat!" or "I'm a vegetarian." I think most of the people I work with and live with—well, certainly the people I live with—understand and accept it. But eating seafood is a compromise and my way of keeping peace with those who have no willingness or idea of how to cook vegan or vegetarian food.

I have certainly reduced my milk and my cheese consumption, but I haven't given it up. And while I'm absolutely abhorrent of factory farming, I suppose in a larger worldview on this subject I tend to agree with Donna Haraway: I wish there were a place for certain farm animals. I'm not sure if there should be a place to *kill* them for us. I'm not sure if I think we should take it upon ourselves, to have the *right* to kill them for food. I'm not sure about that; that's a trickier element. But in an ideal world, where a farm would be giving sustenance, and shelter, and food to animals, I don't really see anything wrong with taking their milk, or making cheese. Of course factory farming is the lion's share; certainly around where I live, there's no local dairy or local farm where I can buy goods where I know the animals have been well raised. But I don't feel that all use is abuse, in fact. I've written a little bit about this, where I've been critical of those who can only see domestic animals as our slaves and believe that they shouldn't exist because of that. I feel that that is sacrificing them in order to cleanse ourselves of our forms of mastery that we disapprove of. And I don't think we should sacrifice them for our feelings of purity. I would hope we could do better than that. And I say that knowing that in many situations, use is abuse, but I don't want to believe it has to be.

SCOTT CAREY: You mentioned that for the most part it has been your research—on horse meat, for example—that has influenced your eating practices. I'm wondering if you have any examples of the opposite, where your dietary practices have impacted your scholarship.

KW: It's been interesting, having written my last book, to go back to my earlier work on horses in nineteenth-century France. I have been struck by the fact that I have a very different consciousness going back to this work, and I have been trying to find ways to make it relevant to the larger ethical questions that I kept coming up against in *Thinking Animals* and that weren't always present in my earlier work. I have been trying to think my way back into how this historical work might be relevant today. So I am trying to ask, and this has come up in the context of a chapter I'm writing on the rise of animal protection, not just in a logical way—in terms of how it comes about—but how do questions of what we would call today "empathy" matter? The term *empathy* didn't exist in the nineteenth century, in any language, so I am asking how questions of pity and of sympathy—were both influential in and inadequate to bringing about positive changes in animal protection. So that's just a long way around to say, yes, my point of view has changed through my work and now is reshaping the work I had done and trying to find a new frame for it. That goes for my dietary practices, as well, because I don't separate my dietary practices from a kind of larger view of how I wish animals were in the world, or domestic animals in particular.

SC: We were wondering if you could speak about the ethical promises and limitations of posthumanism.

KW: With regard to posthumanism, it's funny: I sometimes claim to be a posthumanist, but I'm not sure that I am. There are so many definitions of posthumanism running about. I agree with a posthumanism which argues that we, humans, are not the reason for the world and that decenters the human in order to say that animals, nature, others, are not here for us. Such a posthumanism emphasizes that we are all thrown into the world, are all part of its makeup and its evolution, but that there is not an organizing principle over which we have control, or for which we are the telos. In that regard, I'm a posthumanist. I think we should be wary of the hierarchies that we set up, because they are just that—hierarchies that *we* set up. They are not *in* the world, I don't think.

I suppose the contradiction for me is that I think ethics is a humanist endeavor. I mean yes, Frans de Waal can show that there are certain ethical understandings in other apes, I'm not denying that, but I think ethics as a practice of worldview and as a particular structured viewpoint, is probably a human construct. I don't think there is a single ethical viewpoint. I don't want to be so posthumanist as to dismiss all the ethical constructs we are working with and always having to rethink. I don't want to be so posthumanist as to throw out the kinds of ethical frameworks that humanism has done much to construct. But I want to be posthumanist so that nonhuman animals, and nature, and others are integrated into those ethical constructs, such that humans are not the sole focus, or center. I find some kinds of posthumanist theorizing to be such that they absolve humans of our responsibility for the kinds of agency we do have in the world, agency that has often been destructive. Not all of it destructive—I think we should be able to take credit when credit is due, but we should also take responsibility for much of the destruction, whether it's climate change, or environmental habitats that are destroyed, and so on. There's always, I find, a potential contradiction within a posthumanist ethics, in that it might absolve the human of a certain responsibility by suggesting that we are posthuman in some way.

Then I suppose the other idea that I've been critical of within posthumanism is that it suggests that somehow we can jump out of our humanness. Now I don't claim to know what our humanness is; I think to be human is constantly changing, and we're constantly redefining it. But I think to assume that we can jump out of our human framework and see the world from another point of view—whether it's of a tick, a lion, or a dog—can risk being hubristic. It's like the Leibnizian "view from nowhere": that we can somehow play God, and I think we should be wary of that as well, and so be wary of our limitations as human, that our humanness is a limitation, as much as it might be a skill in some ways. There are skills involved with it; there are also limitations involved with it. I don't think we can escape our humanness, whatever it is.

SC: Could you tell us a story about a time when your dietary practices have been the subject of awkwardness, celebration, or hostility?

KW: Yes, unfortunately. I have been co-sponsoring a summer fellowship in animal studies, and since the beginning we've had all our meals be vegan—and I've been very supportive of that. Two years ago, we were very excited about doing a vegan meal and we had it at my house, and it was all great. Everybody left, except for two fellows, who were the most strident—I hate to use the term—but they were the sort of vegans who give veganism a bad name, because they, I think, checked the fridge to see if there was milk, and then came and attacked me for having milk in the fridge. Maybe not attacked me, but started asking,

"How can you as a feminist do that, how could you do this?" I felt like, I do this fellowship solely out of my interest in and concern for animals, and I'm not the person to be going after. So I felt really deflated and disappointed by that.

I've also been belittled by certain friends for *not* eating meat. There's one in particular—a friend of the family—who likes to have his steak on the barbecue as he gives me the eggplant and will say something about what a nice cow it must have been, with a little wink toward me. I find that offensive. I suppose there are two different sides of the story.

SC: Could you also talk us through your typical approach to buying, or otherwise procuring, food?

KW: Oh boy! I wish I could be more consistent about that. Unfortunately where I live in Middletown, Connecticut, I have a choice between two big, huge grocery stores. Well, there is a very small vegetarian store, where I go on occasion, but it doesn't have much. I love to cook vegetables, so I'll go to the store and just look to see what looks good, and what looks fresh. I try to look at the organic section and decide. I should say, at the vegetarian store, nothing is local either.

In the summer, when I'm in my house in Massachusetts, I have a much better chance of buying vegetables from the farmer directly, or getting eggs from a farmer. But here, although ideally I would like to buy organic and local, that's an impossibility. If it's organic, it's probably traveled from who knows where. So I'm always trying to weigh various considerations: Should I get the local, should I get the organic, should I get this? There are some fairly local fruits and vegetables, so I try to buy those. If I buy eggs, I get cage-free. I realize that's probably just a Band-Aid to attract people like me who think they're doing something better, but I don't know what my other option would be.

I cook for my husband, I cook for my daughter, who has become a vegetarian. I don't really have a plan; I just go and see what seems to appeal to me. I will, on occasion, buy my husband some meat, because he wants that. I figure if he were to buy it himself, he would probably buy some cheap brand that was even worse for the environment and regarding the treatment of the animals; at least I'll buy one that's humanely certified and grass-fed—for whatever difference that might make. So that's, again, where it gets messy, and I kind of hope that nobody sees me buying it. Of course now I'm telling the world that I do. That's the messy part. I don't want my relations with animals to make it look as though I condemn my friends who have other relationships with animals. At the same time, I don't want to hide my preferences and my hopes for what our relationships could be. It's not much of an outline of how I go about things, but it's basically to say that I look to see how many ingredients there are, I look to see that there aren't all kinds of pesticides, I look to see that Monsanto has not been anywhere near anything I buy, and I look for the labels. If I buy seafood, I try and see that it's not been farmed, that it's wild. I also have little faith that these labels mean very much, but I'm kind of stuck with them, I feel.

SK: Thank you so much. We've come to the end of our list of questions; is there anything you would like to add?

KW: No, no, I don't think so. I think the messiness of eating is important, because I think purity is as much of a danger as messiness. Purity and trying to uphold ways of living that can also do damage to our relations with other humans—and appear as if we are condemning them because of our own self-righteousness. Maybe that's a self-justification for my own messiness, but there you have it.

SK: I'm grateful for your eloquent perspective on the messiness of all of this. It turns out that most of our interviewees have been vegan, so I am glad to include your perspective. Most vegans also acknowledge the messiness, but it's still helpful to be reminded of the many considerations and consequences of our choices and practices.

KW: It's not easy. I mean if it were easy to be vegan, I'd be happy to be vegan. Well, actually, I shouldn't say that; I don't know. I'll just add this. Much of my scholarship has been focused on France because I loved France, and part of loving France was loving cheese, and bread, and wine. I see that as a cultural practice and a farming practice that can be connected to a love of a land and a love of the cows. I don't know if I would want to see that as a negative, and wrong. I know it can be; I know it can be harmful and abusive, but perhaps it doesn't have to be. I enjoy being with horses, for instance, and I think they enjoy when I'm with them, and I think they also enjoy the work they do. I may be wrong, but I think there are some animals who like to work. I want to believe we could have a relationship where humans and animals work together, and enjoy being together.

CHAPTER 7

Waking Up

H. Peter Steeves

Philosopher H. Peter Steeves shares here a number of personal narratives, among them his "conversion story" that tells how he changed his dietary habits, as well as the overall course of his life, following a night of solemn reflection in a college dorm room. Deliberate and forthright, Steeves discusses his thoughts on the enduring impact of forging formative relationships with nonhuman animals, grocery shopping and the illusion of culinary choices, and the ways in which food attends to many needs, few of which are caloric. Steeves concludes by contemplating how we can speak from within the very institutions we try to analyze. We interviewed Steeves by telephone on August 16, 2015.

SAMANTHA KING: Since we're interested in how personal and academic interests are mutually constitutive, we'll begin with biography: where you were born and raised; what your formative cultural, intellectual, and political experiences were; how you became an academic; and how you came into the field in which you now work. You could also discuss where you live now, your interests or hobbies, family, and relationships with companion or other animals.

PETER STEEVES: There really is not much that's special about my story, so please just stop me as it gets boring. I grew up in the American Midwest, surrounded by corn, soybeans, and wheat, forests and rocks, and animals. Growing up in a rural environment, with farms on three sides of me, was deeply formative and constitutive of my identity today.

There were many animals in my life. We always had cats in my family, and although I loved them, it was my boyhood dog, Snowball, who was most important for me. We got Snowball when I was five years old when we went to the shelter looking for a rescue dog. She was a medium-to-large size, mixed-breed with long white hair, who was scheduled to be killed the next day. I was drawn to her immediately, and they told my mother and father that she wouldn't be a good dog to choose. Because she had given birth to a feral litter, they thought she was especially "not domesticated" and unfit to be around families and children. But somehow I was able to convince my mother that this was the dog for us—and boy, did she end up being the dog for me. I often tell those who ask that the two most important people in my early life were my great-grandmother and Snowball. And until Snowball died, when I was in high school, she and I were inseparable. I learned more from her than I learned from pretty much anybody.

So there's Snowball, but in rural Ohio there were so many other animals that were important too: turtles, frogs, deer, raccoons, possums, skunks, birds—you name it. And lots of non-animal life, too, all around the fields—soybeans, wheat, and corn—just as far as you could see across the flat land of Ohio. I didn't know at the time that it was industrial agriculture; for me it just seemed like farmland, and it was beautiful. I think of those fields as really important to my upbringing. It was historical Shawnee Indian territory that we lived on, and you would find old arrowheads, especially after the farmers had gone out in the fields and started plowing, and I would pull them up from underneath the earth. I searched for them, then would go to the public library to read as much as I could about the people who had been there before. An early dose of guilt for me.

There was also a Johnny Appleseed tree in front of my middle school that was important to me. I've always had a thing for trees; they're one of my favorite life forms. But this tree—I wish you could see it—it was *beautiful*, and it had a little plaque near it that said something like: "This tree was planted from a seed that came from an apple that came from a tree that was planted by John Chapman—Johnny Appleseed." I used to go visit that tree; Snowball and I would walk there, even before I was in middle school, and just *think* about it all. It, too, anchored me to history in a way. That tree was really important to me.

SK: Where did you go once you finished high school?

PS: The years I was describing—that was not the easiest time. My father left when I was about seven, so I ended up having a lot of responsibility, raising my younger brother and taking care of the family, essentially. We grew up in more or less abject poverty. So they were difficult years, but one of the good things about being responsible at that age—and sort of a logical, stoic soul in those years—was that I was able to fulfill those responsibilities and also have time to think. From a very young age, I was just obsessed—and I don't use that word lightly—with what you might call "the big questions." How did the universe begin? *Why* did it begin? What is time? Why are we here? What does it all *mean*? Growing up, I thought the best way to pursue such questions would be in science. So throughout high school, and briefly beyond, that's what I wanted to do; I actually went off to study theoretical physics for a while.

It wasn't until I was a bit into that program that I became disillusioned by it, realizing it probably wouldn't give me the sorts of answers that I wanted. Foundational questions aren't allowed to be asked in the sciences. Which is not to put down the sciences at all, because if you just go around asking, "But what is time?" you're not going to get any work done in the lab. So you need to have some people who are *not* asking that. But those were the questions I was most interested in, so that was a crisis. I was at Northwestern in the "Integrated Science Program"—an accelerated program where after five-and-a-half years out of high school, you could have your PhD. You basically had no time to take anything other than all of these science classes. It was fine, and there were a lot of good people there, but it also made me realize that there were huge parts of life that I wasn't getting a chance to think about, let alone *do* anything about. Social, political, and ethical philosophy were extremely important to me—and art and literature—and that was all pushed to the side.

So I left that program. It was a major decision because it seemed to me that I didn't have much of a choice about going into the sciences, which is maybe a strange thing to say—especially since I just said that it had been my dream, and it's true, it was my dream—but when I was growing up, it seemed that if you showed any kind of intelligence, you were pushed toward math and science. I was doing well in school, so people would ask, "Do you think you're going to be a doctor or an engineer?" And then the sentence would stop. So the idea of doing something in the humanities never really appeared as a viable option to me. Even the decision to be a physicist, without applying that science knowledge in medicine or engineering, already seemed like a crazy thing.

A short while after I began at Northwestern, I had the realization: I hadn't really *chosen* the direction that I was headed. That direction was failing me both in that it wasn't allowing me to ask the questions I wanted, and it wasn't allowing me to be grounded in the kind of ethical-political practice that I wanted. And so—I realize that this may sound strange—but, one night, I went into my dorm room, locked the door, and told myself in this clichéd, Cartesian way: Tonight I am going to evaluate my life, and I'm going to ask myself, in terms of my actions and practices, what is it that I've really thought about and decided "This is what I want to do," and what is it that I haven't really thought about but I'm just doing passively anyway? I stayed up all night, and a good part into the next day, just sitting there thinking. By late afternoon of that next day I had decided that this school, this life, wasn't the route for me. And I walked to the administration offices and dropped out.

In a sense I remade my life in that one evening. There have been times when I've gone back to that night, realizing that there were things that I still foolishly left out of my evaluation. It would be a couple of years later that I truly started thinking about animals, and how important they were for me, and my own eating practices. But of course that's something I should have thought about that first night, too. I should have thought about what I ate, and I should have become vegetarian earlier.

SK: Could you talk a little more about that transition to vegetarianism, and when you started to ask questions which are often taken for granted in philosophy—questions like "Who is the 'we'?" when we ask, "What do 'we' know?"

PS: It was January of 1987. I took my first class on Eastern philosophy and Eastern religion, which I knew virtually nothing about. I had been reading Western philosophy since elementary school, when I first discovered Plato, but I knew nothing about Eastern philosophy. So I was intrigued by it. I learned that in some areas of the world, ethical vegetarianism is a necessity; I learned about the Jains, as well. And I learned that there was this huge gap in the sorts of knowledge that I had. When I started thinking about why people who had been thinking about the same philosophical issues I was interested in were not eating animals, it hit me that this was another thing that I was doing in my life that was *thoughtless*. Thoughtless in the literal sense—there was no thought going into it. It was a practice that had just been going on passively in the background in my life, with me too unwise to see it. Even when I went through that radical Cartesian night in Evanston it had not occurred to me that *this* was an aspect of life that I had to think about. So foolish! And that realization has been an important moment for me, because I always keep coming back to it, thinking: What else haven't I done or thought about? What else am I still missing?

I decided I should start thinking about eating animals then, and try to come to some logical conclusion about whether this was good or bad ethically. I immediately stopped eating meat—and never started up again. I think that was one of the key moments when I realized that, first, I hadn't been as radical as I had hoped to be when I originally tried to remake myself; and second, in asking the political/ethical questions that were important to me—questions such as, Why am I not doing anything to stop apartheid? or, Why am I participating in a government that is messing with Central America?—which Reagan was doing at the time—and Who gets counted in my ethics? That notion of "we"? I was happily expanding it to include all sorts of humans. But foolishly, it took a little while for me to realize that "we" was still hiding a lot. The first things I started thinking about were animals. When that hit me, I felt so guilty. Just knowing how important animals had been in my life, personally—that was a total blind spot philosophically. That was the first time I truly began thinking deeply about the question of who's in the "we," as you put it.

VICTORIA MILLIOUS: I'm wondering: Have your thoughts and feelings regarding the consumption of animal products changed over the course of your scholarly career? And if so, how?

PS: The short answer would be extremely boring in that they haven't changed massively since that time I "woke up." The only change is that for the last several years I've *tried* to be more vegan. That's actually one of my greatest failings, and something I'm working on—veganism is a project in my life.

Ultimately, I would like to be able to leave the city and have a place out in the country—a little farm where we grow the things we eat and have some animals. Because it seems to me it would be possible to have chickens, for instance, and to know that they're going to give us the gift of eggs, and when they're done with their productivity, we're still going to take care of them, love them, and give them a good home. So that they're outside the cycle of agribusiness, but still able to be in a community with us. It's impossible to argue against the claim that there are massive amounts of torture and sadness being created in order to give us such things as eggs and milk. So the ethics of that seems clear. But I worry that in some vegan literature there seems to be an underlying philosophical ideology that humanity is separate or bad on a fundamental level for all of these other animals, and that we need to leave other forms of life alone completely. It's a weird, almost libertarian strain of "Don't interfere, don't mess with me" rhetoric. To me, that ideology seems somewhat similar to the ideology of the industrial farmer or the hunter who also sees himself as radically separate and different from the animal. So much so that then it's just a question of where that ideology leads, to which practice? If you separate yourself, does it mean that you think about how to hunt animals and kill them? Or does it mean you should just have nothing to do with them? I'd be more interested in the sort of revolutionary future where we could flourish together. I realize that future may be a "future to come" in the Derridean sense—something that's always arriving. But in that future to come, I could imagine that strict veganism would not be necessary—although certainly vegetarianism would *always* be.

A good deal of the ethical work I'm doing these days actually has to do with an ethic that includes not just animals but plants and nonliving things as well. I'm working on a project for an ethics for rocks, for instance. I think they matter. So these are all things that are on my mind, not how to *widen* the sphere of moral import—I don't like that sort of approach or that model of morality—but thinking about what it means to be together and to flourish, to always be checking back with myself about who I am still leaving out of the "we." Where is my blind spot now? And recently, I'm thinking that there's a lot to do there with plants and with nonliving parts of the world, too. Rocks can flourish. Rocks are part of the community. And some of this surely goes back to those arrowheads, as well as the other rocks I admired and collected as a boy.

VM: How has your writing about animals shaped your dietary practices, and how have your dietary practices/identity shaped your writing? I'm also interested in this concept you were just speaking about in terms of "living with" everything around us; is there a relationship between your writing practices and your approach to "living with"?

PS: There's a lot underneath that question. First, there's the question of academic writing. I'm not completely sold that academia is a good thing, or something of which I want to continue being a part. But then there's the question of just writing in general, which I used to prioritize over other activities. I would put so much effort into writing, hoping maybe it would *be* something, maybe it would make a difference somehow. I was seduced by the idea that it's more public, or it has a greater temporality; it might last beyond me. And that's something I'm trying to get over. Because although I submitted two books to publishers this summer, otherwise I haven't published a book for about eight years now. In that time I purposefully chose not to do as much writing and instead put my time into "lecture performances." They're a mixture of a scholarly lecture, but with live performative elements—dance, theater, magic, music—scattered throughout. I've been doing a lot of those, as well as some painting and installation art. I've been lucky enough to have been asked to create pieces for some galleries, museums, and cultural centers. And all of that—especially the lecture performances, which take a huge amount of time and energy—is so ethereal, ephemeral. They're just there for a moment, and then they're gone. In the beginning I wanted to videotape them to have a record, and now I've stopped even doing that. Because it seems the lure of writing is still always behind the impulse, "Maybe this could speak to somebody who isn't here right now—let's save it." There's too much ego invested in that, which I'm always trying my best to check. So writing is okay—maybe—but I'm trying to see it as not more important than anything else.

This is all a roundabout way of getting to the second part of your question about writing and other practices in life. What I'm trying to do is to see them as not separate. To write about something, and then do something that's informed by the same values or the same kind of thinking as that writing, I see as not being different. I'm trying to break down that distinction.

VM: We're definitely interested in breaking down those distinctions, between writing and eating practices, for example. Most of the editorial team came to this work in part through our own vegetarianism, and the project has certainly prompted us to reflect on these practices, so it's a very personal undertaking.

SK: Yes, I've been a vegetarian for thirty-two years, but I would say—like Victoria—I was fairly thoughtless about it until recently. I was thirteen. I didn't like the idea of eating flesh, and I didn't enjoy it very much, so it didn't feel like a significant sacrifice, if I'm honest. And then over time, I began thinking about it more. One of the big moments for me was reading Jonathan Safran Foer's *Eating Animals*—it became a sort of vehicle for asking deeper questions. His point that we all know about factory farms but don't actually change our practices in response struck a chord.

PS: That's fascinating. Wouldn't you say it's something many people have in common, this thoughtlessness? And then when you're confronted with the thoughtlessness, you end up making a decision? And you could be led there by a book; it could be an experience—but it's something that makes you realize that you've been doing something, not really choosing it, but just doing it.

Although as you say, even after most people know where their meat comes from, they don't change their practices. But to "know" something—there's so much hidden in that word: you can know something and then essentially not know it tomorrow, because there are too many structures in our society trying to make you forget. Until those structures are changed, it's hard to imagine most people choosing the right thing, which I think they would do if those structures were not there. That's one reason I think we need revolution and not just another academic, such as myself, writing an essay about animality. Perhaps the latter can help hasten the former. *Perhaps.*

It's also not a question of showing someone a rational argument. Sometimes that works, but the way we relate to animals is not just a question of doing something illogical. For instance, I have a friend—a fine and smart colleague—whose project once involved race issues. I spoke to her years ago about how she was going to write this book that she thought would be a first step in solving all racism, and she explained that she was going to suggest that most people who are racist in their lives are that way because of *bad induction.* That is, they'd seen a lot of "black people" on TV or in the movies who are criminals, or they were mugged once by a black person, and then they drew the general conclusion that black people are bad. So if they could just be shown that that's a poor inductive sampling technique, then they would not be racist. She had the best intentions, and the best analytic philosophy training, but I remember thinking: This is utterly misunderstanding everything! Racism is not a problem of poor logical thinking. Racism is a question of structures, of ideology, of history and historical materialism—massive forces that get us to live and inhabit the world in a particular way.

And I think the same sort of thing is going on with animals. We can show people and tell them; we can even make slaughterhouses with glass walls. But unless you address the structures in a much deeper ideological way, it's harder to get most people to see what's right in front of them. So it's great you're doing this project to see how some people do come to that realization, because maybe there will be something to be gained by thinking about what they have in common, or the particular journeys and stories that people have, in terms of then making it more successful for others, too.

VM: Can you tell us a story about a time when your vegetarianism has been the subject of awkwardness?

PS: In terms of the social aspect of dining, things have been the most difficult for me because I don't *drink*. Imagine: an academic who doesn't drink alcohol!

What has been interesting, at least for me, is that two years ago I became the director of the Humanities Center at DePaul. I agreed to do it on the condition that I could create events that were not just the same old stuff we tend to do in academia. I've really enjoyed it; it's a creative freedom that I feel honored to have. At the Center we have a lot of special guests coming in, sometimes really big names. And after these events, we go out to eat. In the past, the Center often didn't have a speakers' dinner; events were catered so as to have appetizers and the like for the public at a reception. At DePaul, as with most universities, we're forced to use the on-campus caterer, and it's just the worst—and expensive, too. So the first thing I said was, We're going to do a lot less of that. Because looking over their menu, if it wasn't just a plate of carrot sticks, then the majority of stuff had either dead animals or animal products in it. It just seemed like throwing money away for the carrots, and there were no great options for vegans or vegetarians. So I ended up taking our guests out to dinner at a nice restaurant, a Turkish place. I could preorder a lot of dishes before we arrived, dishes without animal products. And then I could tell people it was all going to be served family-style to take the stress out of ordering. Which is true, but the main reason I did it was for this ethical reason. I made a commitment to myself that this was one tiny area within academia where I actually have a bit of power, so I would never sign off on spending money on something I thought was unethical, such as serving animal products.

And that has been challenging at times. Because the question of hospitality can conflict with my strong moral commitment against spending university funds over which I'm in charge to promote death and suffering. There have been a couple of times when people have asked, "Where is the meat?" It's been awkward, but I've tried to think of it as a "teachable moment." I don't want to have the relationship with a guest that I'm somehow teaching them and instructing them—even if we admit that's what we're always doing for each other; we're always acting as examples. But I do use it as a moment to explain why there is no meat there. And there have only been two times when someone became so distressed by the lack of meat that they demanded some. I hope it was the right decision that both times I didn't pay for that added meat. And we at least talked about it a bit. That's an uncomfortable conversation to have because it's hard not to feel superior, or to talk down to someone—both things I just loathe. I'm not much of an outward proselytizer; I just try to be an example. So there have been awkward discussions, but I hope they ended up being worthwhile.

ISABEL MACQUARRIE: Changing gears a bit, in terms of personal grocery shopping, we're often overwhelmed with choice, for example organic, vegan, local, etc. Can you talk us through what goes through your head as you're grocery shopping? Do you see it as an ethical endeavor?

PS: This is a really hard one for me. I think it's important, obviously, to be vegetarian and I try—even as I fail—to be vegan. In terms of local, we have a few good places in Chicago, so we try to do that as much as we can, but it's not enough. I don't know if it's completely rational, but I harbor a hatred for Whole Foods. The guy [John Mackey] who runs Whole Foods is a kind of libertarian who hates co-ops and has fought against them and other workers' movements. And he's a vegan himself, but he says, "Well, the dollar demands it, so I'll sell meat." What are you supporting when you support that? But then what are the options? It you don't support that, you just support another huge chain probably. It's a difficult thing in the sense that there are a lot of options, but not *really* very many. The many-options thing always seems to me to be one of the ways in which liberalism and capitalism try to fake us out, to make us think we're free.

I remember about twenty years ago I had friends visit the U.S. from Latin America—a couple of whom had never been in a huge supermarket before. And one of the guys, I lost him in the cereal aisle, because he just stood there. The entire aisle was full of different brands—probably more than a hundred choices of cereal—and he was overwhelmed to the point of not being able to choose. You know the old story about Thales's mule? He had two bales of hay in front of him and he tried to decide which one to start eating first, but they looked exactly the same, so he couldn't make a decision and he starved to death. Sort of like what was happening with my friend here; there was just no way to choose. And I remember that the upshot was, "Wow, you have so much freedom here!" This is one of the tricks that capitalism plays on us to make us think we're free—when freedom gets equated with the number of consumer choices on a shelf. Unless you're producing your own food, you are supporting systems that are inherently horrible for fostering real community and ethical living.

In terms of food more generally, it's something with which I've always struggled. Grocery shopping is a very intimate thing. I often fail miserably on many fronts. As I reflect on the ways in which my eating practices developed, I can find little moments that were critical. In my essay on fake meat, I tell the story of my grandfather's telling me when I'm seven that I'm the man of the house now, so I have to eat steak for the first time. There are these moments—Freudian or otherwise—where you say: That was surely informative of the ways in which I relate to food today.

There is one other childhood moment in particular. It was just before Easter when I was about ten years old. It was a particularly difficult time because my grandfather had recently died, and my father had left a couple of years earlier, so—living under the patriarchy—we didn't have much support. But we were at a restaurant on the invitation of the owner, and it was a time in my life when we weren't eating every day, so we were really glad to be there. Even though I was glad about that, I was pretty sad in general. The restaurant was owned by a guy named Joey, who had been friends with my grandfather. He was a great guy and he saw that I was feeling really bad, so he took me into the basement and there were these cakes shaped like lambs—Easter cakes—on a long table. Some of them were decorated with white icing and toasted coconut to make the "wool." Some weren't decorated at all. But all of them were broken, so he wasn't going to be able to use or sell them. He gave me a fork and said, "Here, eat all you want, because when you have something delicious to eat you always feel better about everything." And given the horrible things going on in my life—and then to be told, "If you have emotional problems, eat"—this was, unfortunately, seminal for me. And interestingly, in reflection, it's tied to a kind of animality, as well. Lamb cakes.

So those are some things that are on my mind when I'm shopping: Am I shopping for something that is ethical? I hope so. For the people that made it, for the world, and for me. But food fulfills a lot of needs that we have, and sometimes it's not really caloric needs.

IM: Could you tell us about the reasoning behind including recipes in some of your scholarly articles?

PS: The reason for the recipes is twofold. First, I'm always interested in borders and boundaries, and conceptual ways of categorizing the world—what work are those borders, boundaries, and concepts doing? Because once they're in place, they do the thinking for us—and that's always dangerous. So I try to find ways to question those borders and boundaries. In my lecture performances, for example, I'll be talking about a passage from Husserl one moment, and then suddenly singing a song by David Bowie. There are multiple ways to make a point, and sometimes art is the *best* way. Additionally, I don't see why we have to think that writing is different from acting, that doing is different from thinking, that giving a lecture is different from singing. So I included the recipes to mess with the expectations of what a scholarly article should be. I hoped that people would read through the recipes like the sentences of the essay itself, and then think, why does this seem out of place? And is it really out of place?

The second reason I decided to put them in, and the reason I decided to go for recipes *per se*, is because recipes are a kind of instruction—they demand through an imperative. The moment you start reading, they say, "Take this, mix this, bake this." There's something interesting about that, because there's a kind of command to a book, too—a command to all writing. When we write, even if we don't come out and say, "Now do this!" we are making imperatives, we are saying, "I think this is right, I think this is true, and I hope you will too." It's hard to have a conversation in writing—as Plato knew well—but you're always wishing for it. So I thought including the recipe, a kind of writing that is an imperative, would be a nice way of saying that the other writing in this essay—the discussion of fake meat and phenomenology—that's an imperative too. I wanted to juxtapose those two things without mentioning having done so, hoping they would have a force to them.

IM: I have a follow-up to your comment that you're considering leaving academia. I'm wondering who—or which institutions—do you see as having a responsibility to help people challenge the natural attitude toward our relation with animals, and the structures which perpetuate this attitude?

PS: That's an important conceptual question, but it also has real implications for how the future might unfold. And I can say up front that I don't have a great answer. *Every* institution should be ethical, yet as an anarchist, I don't really think that there should be any institutions around us at all.

It seems to me, anyhow, that I am an academic somewhat by accident. I have been in school my whole life, and I've personally liked school. But I don't know necessarily that it's the thing that, ethically, I'm supposed to be doing with my time. And I have my fears that schools are actually immoral institutions. So I'm not sure that I will stay, that's true. At least I don't see the university as necessarily being a part of the solution.

Schooling as an institution in general has a pretty sordid history. You cannot separate the educational system from the prison system—in the U.S. at least. So it does seem to me that the university represents the pinnacle of all that is wrong with education in general. It costs a huge amount of money, you train for a job, and you are—for the most part—forced to go into it out of fear. A good number of my students are afraid of what would happen if they dropped out of school: They think they wouldn't get a lucrative career, they wouldn't know what to do in life. So sometimes we talk together about this and I suggest maybe you should be afraid of what will happen if you *don't* drop out. If you stick it out for all these years, what sort of person will it turn you into? That's a complicated discussion to have, in that you are speaking from within the institution you are trying to analyze. And I always try to be aware of the fact that I have been given the power to do that, and that there is a necessary privilege and a power relationship in the classroom, even when I'm trying to undermine it. My wife, no doubt, is better at this in her classes than I am. She recently got a *tattoo* spouting anarchic phenomenological communitarianism, a theory for which I've been arguing for more than twenty years. She's hardcore. I'm still a bit of a timid Midwestern forest creature when it comes to confrontation.

The short version of this would be to say that when the revolution comes, it's not going to come from within schools. And after the revolution comes, there probably won't *be* schools. It's impossible to separate universities from capitalism. And it's impossible to separate the university from an Enlightenment ideal of what a human being is. And that Enlightenment ideal implies liberalism—big "L" Liberalism, as in Hobbes, Locke, Rousseau, Descartes—which is all part of the problem. "I'm an isolated individual; we all come from a state of nature; we're only here because of a social contract; reason will shine its light on everything; there's progress such that you can know more than you knew yesterday, and that makes you a better person, and all of that can be commodified and turned into a dollar amount such that you can then sell your labor on the market." All of this comes as a package deal. So it seems to me there's nothing you can learn from someone who's being paid to tell you, in a place you're compelled to go, that could not be taught to you better by someone who isn't being paid, in a completely different context. My wife, again, is really good on this subject. Her current project is child liberation—another constituent in the "we" I came to realize embarrassingly late in the game. I can definitely say that as an anarchist, I think we can reap all the benefits of education without the educational *institutions* we have right now.

SK: Is there anything you'd like to add before we conclude?

PS: There is one thing that I think runs throughout much of what we've been discussing but hasn't been thematized specifically. And that is to put ethics before ontology. It is a rule I'm trying to live by. I'm so trained as an intellectual, as a philosopher, to think along the lines of, "What is x?"—to do a kind of metaphysics first—but it has become utterly clear to me that this is another way of derailing the most important conversations that we can have.

Take, for example, a question about our relationship to domestic animals or companion animals, and the roles they play in our life. The history of Western thought teaches us that we first must think about the ontological categories that are at work, asking what *is* a companion animal, and is a companion animal different than a domesticated or wild animal, for instance—does it have an essence? But before engaging in this ontology, I would want to ask why might somebody think that such categories are worthwhile to have in the first place? If we accept those categories as worthy of investigation, what do we want out of them? What work are the categories doing for us, and is it good or bad work?

I'll give you one example; it's an example that I say hesitantly, because if someone hears it out of context, it could sound really horrible. But suppose somebody came up to me and said, "So, I was wondering, do you think that we could fairly categorize the world into Jews and non-Jews?" Now, the philosopher's cold response would be to think: "Hmm, is there something that *is* Jewishness? Is there an *eidos*, an essence to it? And is it constitutive of identity on a metaphysical basis? Is this identity/essence just culturally informed? Is it historically informed?" etc. Perhaps all of these questions are abstractly interesting. But I hope that if someone came up to me and said, "Do you think the world can be divided up into Jews and non-Jews?" my first response would not be a metaphysical one. It would, instead, be, "Wait. Why are you asking this? What in the world is behind that question?! What would you do with an answer if it were given? I'm really worried about this!" This is an *ethical* project before any ontology ever gets off the ground. And, frankly, it should scare us if somebody asks that question.

That's a visceral example: a moment where you can see that before we start doing metaphysics, we should think about the ethics behind asking the question. For me, though, this is always the case with *every* question. When we ask, "Is that person straight, gay, ill, an adult, disabled, a criminal, a woman, an American, a human, etc.?"—or, more to the point of the project at hand, when we ask, "Do you eat organic or not organic? Do you eat fake meat? Are you a vegan, a vegetarian, an omnivore, etc.?"—our tendency is to try to define what each of these terms or concepts is, find their ontological essence, and then ask whether they're real or not. But I find it much more instructive to interrogate the question itself, and to ask about the ethics and politics of the question and the concepts before doing the ontology.

Now take something like anthropomorphism—which actually seems like a fake problem to me because it assumes that I know what it is to be a human and then unfairly transfer that sense to another creature. I don't know what it is to be a human without also knowing what it is to be something else. I would not know what love is if I had not had my particular relationship with Snowball. I would not know what loyalty or trust is if I hadn't seen that in Snowball before I ever "gave it back" to her. There's no way I could have a human's notion of these ideas before I am constituted in the more-than-human world, so anthropomorphism is a nonproblem for me. But even deeper than that, anthropomorphism assumes that there's something it is to *be* human, an *eidos*, and, conversely, there's something it is to be *not* human such that one might ask, "Are you falsely attributing a human quality to a nonhuman?" The ethics-before-ontology approach would say, "Why is *that* question being asked?" It's similar, for instance, to the criminality question. Rather than ask, "How should we define 'a criminal'?" I would suggest we investigate why anyone is interested in dividing the world up into criminals and noncriminals—what social, political, ethical work does that do once you have such a concept, no matter how it's defined? What does it assume and make possible? I would say, then, that rather than ask, "Did you anthropomorphize that nonhuman animal when you said she 'loves' her young?" or, more generally, "Should we treat animals equally compared to humans?" I am less interested in investigating the biological, metaphysical, ontological nature of the concepts of "animal" and "human" and instead suggest we start the discussion by finding out why *this* is the question, why someone might think that "human" is a good category to have, and what might be done in the world—socially, politically, ethically—with such a category once we're done philosophizing. "Vegetarianism" is an ethical commitment as a concept before it is a practice determining what goes in our mouths. But "human" is also an ethical commitment long before it's a biological or metaphysical one. So is "animal." So is "young." So is "love." So is *everything*. Eating is messy, it's true. As are thinking and being.

CHAPTER 8

Entangled

María Elena García

A professor of the Comparative History of Ideas, María Elena García uses questions about the animal to expand existing intersectional research on labor, the environment, and race, with specific attention to how violence manifests at the level of the body. Presently studying the cultural capital of the Peruvian guinea pig, an animal with a strong indigenous legacy, García does work that is as much about settler colonial politics and multispecies relations as it is about Peru's contemporary high-end fusion cuisine. In her interview García shares her abrupt transition to veganism, her thoughts on the additional lenses motherhood offers to research, and the challenges of weaving ethical commitments with familial traditions. We interviewed García in Edmonton, Alberta, on June 21, 2016.

SAMANTHA KING: Could we start with biography—where you were born and raised; your formative political, intellectual, cultural experiences; and how you came into the work you do?

MARÍA ELENA GARCÍA: I was born in Lima, Peru, in 1971, and was there until 1976, when my family and I had to leave the country. We lived in Venezuela for about a year, then in Puerto Rico for another year, and finally we made our way to Mexico City, where we stayed for about six years. Mexico was really important for me. It was the longest I had ever lived in one place; I was there from when I was eight to fourteen—formative years. I loved it; it was home. In 1985 my father, who had been working as an engineer for different telecommunications companies, came home one day and said, "We are moving to the United States in two weeks." So in two weeks we moved from Mexico—our home—to Virginia. I mention that particular move for a few reasons. One, going to the United States as a Peruvian girl was particularly intense. Today, there are quite large Peruvian, Bolivian, Salvadoran, and Latinx communities in general in Virginia, but not at that time. While living in Puerto Rico was technically living in the U.S., it felt more like being in Latin America. And in Mexico, we were Peruvian but also part of the Latin American community. But starting high school in Virginia and facing people asking me, "Did you go to school on a donkey? Have you ever seen a building before? Did you live in grass huts?"—those kinds of things had a huge impact on me. I tell my students that I became Peruvian in Virginia because it forced me to reflect critically on my positionality: "Who am I? What are these labels people are throwing at me? Why are they asking me to label myself?"

This move also represented a radical shift because in the past we would fly to Lima every summer. My grandmother and I were especially close. She passed away a couple of years ago, but she was probably the most important person in my life. Going back to see her, my aunts and uncles, and my cousins was really important. I am not sure I knew how important it was until we could no longer travel there. That year, 1985, was when violence in Lima escalated because of the war between the state and the Shining Path—the Maoist organization that declared war on the Peruvian government in 1980. So in 1985 we left Mexico, arrived in Virginia, and my parents announced that we would not be returning to Peru until the violence ended. It was a very disruptive moment, and I think it had a huge impact on who I am. I wanted to know more about this violence, about why we couldn't go back to Peru and what was happening there. As I learned more, especially about the disproportionate impact of this violence on Indigenous people, I became obsessed with my birth country. In one phone conversation with my grandmother, I learned that my grandfather was a Quechua miner from Ayacucho, something I had never heard until then. All these things came together in a particular way for me and eventually led me to think about anthropology. Even going to graduate school was something I never anticipated. I was the first woman in my family to go to grad school, and that was thanks to a great friend who was an archaeology professor in college who pushed me to think about that. Because of the expected year conducting fieldwork, I also saw anthropology as a discipline that would allow me to go home. Naïvely, I thought I could go home and figure out a way to work to prevent this kind of violence from happening.

SK: And when you were in university, did that coincide with the critiques within anthropology about colonial legacies?

MEG: Yes, it did. Some of the courses I took perpetuated essentialist representations of "others" and racist stereotypes—these are exactly the kinds of courses I teach against now. But there were a few people who were beginning to raise questions and consider decolonial approaches to teaching and research. Because of my own background, the questions I was asking resonated very much with those critiques. During my first years in grad school I began to think critically about this and read more about what had been written at that point. Some of my professors championed my own position and research in Peru as representing an "authentic voice," while others discounted it as too subjective to be "science."

SK: So how did you come to write about animals?

MEG: When I started grad school in '94, my work was focused on Indigenous cultural politics and Indigenous political mobilization in Peru. In the late 1990s and early 2000s, the scholarship on Indigenous politics in Latin America was focused on the question of why nothing was happening in Peru—a country with a large Indigenous population—as opposed to Bolivia or Mexico where you saw massive political mobilizations of Indigenous organizations at national scales. In my work I argued against that analysis and explored instead what was actually happening at local, regional, and national scales in Peru. I developed this work during a two-year postdoctoral fellowship at Wesleyan University, and then found my way to Sarah Lawrence College [SLC] as an assistant professor of anthropology.

At SLC, I taught classes on indigeneity, human rights, and violence in Latin America. Because SLC was so small and offered a lot of freedom in terms of what one could teach, it was also very open to different topics and approaches. One of my colleagues there, Karen Rader, had written a terrific book called *Making Mice*. She was giving a talk about this work and had shared chapters from her book which I read before attending the lecture. I was struck by the fact that while the book was about the making of the mouse as a lab animal, there was no real mention of the actual mouse as a living, breathing being that feels pain. I asked Karen what I thought was an innocent question about this, but it led to a long argument which ended with her saying, "I'm done having this conversation. If you care so much about animals, why don't you teach a class on animal rights?" I remember laughing at this and thinking, "That's not the point." But I could not stop thinking about Karen's suggestion.

As an anthropologist in a very human-centric discipline I wondered, What would an anthropology of animal rights look like? I went to the public library to see what I could find in animal studies, and I was blown away by the range of texts I found: books by philosophers, feminists, historians, and literary critics, all writing about animals from multiple vantage points. This led me to develop a class on animals, which was really about comparative social movements. I was interested in the ways considering animals might expand our thinking about labor, the environment, gender, and race.

I was just beginning to learn about animal studies, but animals had been an important part of my life. My mother had studied to be a vet and while in Virginia worked with rescue organizations and fostered several animals. Companion animals had always been very much a part of our lives—but we also ate everything. We came to this country and I remember we were forbidden from eating in the cafeteria; we had to go home to eat Peruvian food. The first few years one of our typical dishes was a big bowl of chicken hearts, sautéed in garlic and butter, because it was cheap. People talk about this all the time: the disconnect between loving and eating animals. But it was teaching this first class on animals at SLC that got me thinking more explicitly about this disconnect.

That class was transformative for me. I had thirteen students, some of whom I am still very close with. I had a handful of animal rescuers, vegetarians, and vegans. But primarily I had students with very specific motivations: One young woman was there to defend her right to hunt, and another African American woman from Georgia wanted to defend the Christian notion of dominion. It was a fantastic group of students and they each brought such different perspectives. I walked into that class saying, "This is not my field, so let's do this together." I thought I was going to push them to think about race, class, gender, but they pushed *me* to take animals seriously. I always use a lot of audiovisual materials in my classes. Witnessing violence against animals is something I had previously avoided at all costs. But in this class, I could not ask my students to look and listen if I was not going to do that too. I remember waiting until the last possible minute to watch documentaries depicting horrible violence, just so I could be prepared to watch them again with my students in class. Some students brought their companion animals: dogs, cats, rabbits, and even a python. The seminar, then, became a transformative space. Toward the end of that class, I made a radical break and turned toward a vegan diet. While my approach at the time was not great—I was exhausted, with no time, meaning we ate only pasta, beans, and rice—I think of that as a formative moment, and the beginning of an intense journey around food, ethics, and alternative relations with other animals.

SK: Did you then come to do research on animals because you started seeing things differently as a result of teaching this class, or did you have an intention that animals were going to become a focus?

MEG: Actually, I never thought it would be part of my research. I thought I would only teach one class on the subject. The funny thing is, I have always seen the personal, political, and intellectual as intertwined in my life, but in this moment I made this artificial separation where I would somehow *only* teach about animals. But as that seminar progressed, I started to wonder what I had missed in my previous research. There are important relationships between many of the Indigenous communities and activists with whom I've worked, and, in particular, guinea pigs, alpacas, dogs, and other animals, but I never really thought about it. It was through teaching—and reading work by other scholars who were thinking with animals—that I became interested in thinking more broadly about these kinds of connections. This also coincided with the Peruvian "gastronomic revolution" that I am now writing about. Today my work certainly informs my teaching, and vice versa.

SK: Could you talk about the potential you see in putting postcolonial studies and animal studies in conversation with each other?

MEG: Animal studies is an important interdisciplinary field. It's pushing the humanities, social sciences, and natural sciences in significant ways, but there are blind spots. One of the concerns I have with animal studies is—in some cases—the lack of openness to thinking seriously about race, colonial legacies, or indigeneity. This is changing, but there is still work to be done in bringing these concerns together. Similarly, it is important to think about the consequences of lack of engagement with nonhuman animals in the work of postcolonial scholars and critical race theorists, for instance. This is challenging, but we also need to consider the impact of colonialism and imperialism on more than human beings.

One of the things I struggle with in my work is that in some ways, bringing in the nonhuman—whether an animal or a mountain—can be seen as reproducing postcolonial violence: Placing *animal* and *human* next to one another in certain contexts, the so-called "equating" of human and animal life, can be perceived as reproducing naturalized hierarchies that place certain animals above certain humans. How do you begin to question this binary when some marginalized communities are trying to fight for the status of humans to begin with, and when it's wrapped up in broader struggles for human rights and justice? At the University of Washington, the Animal Studies Working Group is moving toward what some are calling "intersectional animal studies": putting animal studies in conversation with postcolonial studies, critical race theorists, queer theorists, and others. We consider the mattering of nonhuman animals in a way that intersects with human concerns around difference, inequality, and justice.

It's tricky, but when you begin to think about the impact of colonialism on bodies and life, human or nonhuman, and the ways in which we're always entangled, some amazing things can happen. It also reminds us—and Claire Kim's latest book is a terrific example of this—of the fact that race and nature have always been intimately linked; you can't really disentangle them.

SK: Maybe I will jump to ask you about that—how do you navigate the danger, in certain situations, of making "equations" between humans and animals?

MEG: Even if we very carefully avoid the notion of equating, work placing humans and other animals in the same analytic frame can still be perceived that way. So we have to be extremely thoughtful and careful about understanding particular histories, and particular contexts or moments. Inspired by Chad Allen's theoretical and methodological approach to global Native literary studies, I find the term *critical juxtaposition* quite useful in this. I'm trying to think through critically juxtaposing the impact of violence on particular lives and bodies. For example, a lot of the work I do is embedded in thinking about the history of political violence in Peru. I run study abroad programs to Peru designed to have students think about this particular history and its legacies; we work with artists and others who are trying to keep alive discussions of colonial legacies, political memory, and contemporary violence around extraction, environmental justice, and more. It's a central part of what I do, but one of the things I've been thinking a lot about is how nonhuman others fit into this work. I have been struck by the fact that animals don't really figure in narratives of political violence, except for a couple of iconic cases. The Shining Path introduced itself to Lima by hanging dead dogs on lampposts. There is one photograph of one of these dogs that is one of the iconic images of this moment. People say, "That's horrible," but the sentiment does not necessarily translate into concern for other dogs or even for Indigenous victims of the violence. There is another emblematic moment that is in the memory of a lot of people: the Shining Path massacre of hundreds of alpacas. The alpacas were part of a U.S.-backed agronomy institute, so they were seen as part of U.S. imperial projects by the Shining Path and recognized as victims by the state and other Peruvians. I have been thinking lately about how we might think more about the impact of political violence on both humans and nonhumans. Many of the testimonies of Indigenous Peruvians include the theft or mutilation or killing of their animals as violence they experienced. I want to go back through those testimonies, to track the animals throughout, and to talk with Native and *campesino* activists about this project and approach.

SK: So it is in there; it's just that it has not been talked about or recognized.

MEG: Right, it is in there, and recognized only rarely. But doing this kind of work is challenging for the reasons we have discussed. No matter how careful I am, this kind of project can be seen as equating human and nonhuman deaths and minimizing the suffering of human beings. This is something I am already facing in my work on guinea pigs. Given how charged discussions still are today about what happened during the war and who is to blame, developing this project will be extremely hard, and I will have to be very careful. The last thing I would want is to be perceived as reproducing ongoing violence against marginalized communities in Peru—especially given the histories of animalization of Indigenous peoples and others.

SK: I was going to ask you about the significance of the guinea pigs as a focus of study and the ethical and political challenges of doing this kind of work. Is there anything you want to add to what you have said already?

MEG: Guinea pigs are a fascinating entry point into the complexity of contemporary Peru. It's an iconic animal; tourists who travel to Peru—even before the food boom—will often try guinea pig. You can find tons of Google images of people posing with guinea pig dishes, saying "Look, I'm eating guinea pig!" Because they are considered food animals in Peru but not in the U.S., consuming guinea pig becomes part of the exotic adventure. It's a way to consume the other.

Guinea pigs also have a significant place in Indigenous and Andean history. They were domesticated centuries ago by Native peoples; they live with people in their homes, and Indigenous women have particularly strong relationships to them as they have traditionally been the ones to care for—and kill—guinea pigs. But in this moment of gastronomic high-end fusion cuisine, there is a need to authenticate culinary fusion as Peruvian, and the guinea pig has become important in this move. The representational dimension of this is also interesting because it allows us to talk about how the figure of the guinea pig in television commercials, in restaurant posters, and even as presidential mascots can signal colonial histories and legacies, and the racialization of certain people.

I'm also concerned about actual guinea pigs—the animals themselves. In recent years, a guinea pig boom has accompanied the Peruvian gastronomic revolution. There has been a push to industrialize guinea pig production, develop the export market, to genetically "improve" these animals, as part of the move toward national "progress" and modernity. Part of my current project includes tracking the guinea pig, the impact of this moment on guinea pig bodies, and the consequences it has on relations between guinea pigs and Indigenous and *campesino* farmers. The leading guinea pig expert in the country says the animal is going the way of the chicken—following a model we know is horrific. So, while the guinea pig is a fascinating entry point to think about race, species, and capital in Peru, it is sometimes difficult to talk about their centrality to my work.

I will give you one example. I was at a guinea pig breeding farm a few years ago; I was pregnant at the time, and for me that created an even more intense connection with the 1,400 or so pregnant guinea pigs I saw at the farm. I remember being really shaken up by this encounter. I returned to my grandmother's apartment in Lima where my mom was also staying, and tried talking with her about the experience; about my relating to these animals in part *because* we shared the condition of pregnancy. She was furious with me. In her view, I was comparing my human body, and perhaps my child, to animals. And for her, this was the worst sign of disrespect to women. It was a lesson for me: I don't talk to my mother about these things anymore. But it was also a really important moment for me to think more carefully about my approach, my methods, and to anticipate critiques, but without shying away from thinking with and about animals as sentient beings.

SK: It is a courageous project in many ways. One thing that's been coming up in our interviews is how social science and humanities scholars are being more influenced by people who do work on animal sentience and emotion. Is that true for you?

MEG: Yes. I really want to dive into the work ethologists have conducted on guinea pigs and other animals. I was just at the Race and Animals Institute at Wesleyan and this was part of a really interesting conversation about foregrounding the "real-life" guinea pig more fully. I've been thinking a lot about ways to include the sensorial dimension of this work, and engaging with the work around animal sentience, and bringing that in as a way to explore the rich and complex emotional lives of animals. Spending some time with people who live and work closely with guinea pigs will also be helpful in this regard.

SK: That sounds fantastic. I have one more question in terms of theory and scholarship: Is there a key concept that guides your work in this area?

MEG: I've thought a lot about that question. I would say that *entanglement* is something I keep going back to because I find it really helpful in teaching about these issues. It seems to be something that students can latch on to that's not threatening, strangely, and that allows them to see new connections and possibilities. We tend to think so much about the differences between humans and nonhumans—and that's important—but finding ways of talking about shared precarity and vulnerability can open us up to compassion in ways that move us away from the common question, "Why should we care about animals when there is still so much human suffering?"

SK: Can you describe your current relationship to consuming meat and other animal products?

MEG: I would say it's messy. But I live in Seattle, which I am grateful for because it makes it easy for us to essentially live a vegan lifestyle when I'm at home. That said, I do not consider myself a vegan and prefer not to label myself that way. In addition to my work in Peru, I also work closely with Native scholars and activists in the Pacific Northwest. At the University of Washington, as I mentioned earlier, we are trying to build an "intersectional" animal studies, but we are also working hard to build global Indigenous studies on campus. Over the past couple of years, there has been great momentum around that with new hires and the building of wəɫəbʔaltxʷ, which in the Lushootseed language of the Coast Salish peoples means Intellectual House, a longhouse-style facility that serves as a center for Indigenous students and as a reminder to all who see it that we are on Native lands. This means we find ourselves invited to events on Native sovereignty, food sovereignty, and to ceremonial spaces, and there are moments when communities will offer salmon, for example. For me, in those moments it is extremely important to accept these invitations and to eat with others. This is similar in Peru. When I was starting to think about these issues and when we had that drastic move toward veganism in the U.S., I would still go home to Peru and eat meat. At my goddaughter's home, for example, my *compadres* would offer a whole guinea pig and I would always eat it. These are important moments of relationality, solidarity, and affection. Or with my grandmother, she and I would have intense conversations where I would tell her what I was teaching and thinking about, and why I wasn't eating meat, and she would say, "That's great! Now here is your goat, so eat." Food is love and relation and culture, and for all these complicated reasons, in certain spaces I eat more flexibly. This got harder and harder as my work and teaching developed, and as my work began to focus more on guinea pigs. But I also figured out ways to navigate and negotiate these contradictions. With my parents in Virginia it depends on the moment. My dad actually went vegan for a while. He watched *Forks over Knives*, he read *The China Diet*, and for health reasons started thinking more about his patterns of consumption. For my mother these discussions are interestingly gendered, and so for a long time she felt I deprived my partner by not cooking meat for him.

SK: Has your partner been happy to go along with your newfound veganism?

MEG: He has really been incredibly supportive, wonderfully supportive. Initially we did have a lot of conversations. The fights began over milk for some reason. He'd say, "I understand you don't want to do this but I really want to drink this," or "How about maybe I have some chicken?" My response to that was, "Okay, that's fine. But you have to buy it and prepare it because I'm not going to do it." And he just didn't. It's harder for him when we go to see his family in El Paso, who have been less open. They have seen me as depriving him, and not being a good wife, though this seems to have lessened. I would say that we are now both on the same page, that we both see our relationship to food and eating as part of a long and winding journey.

SK: Would you like to add anything about the relationship between your research interests and your writing and your dietary practices, and how that relationship might have changed over time?

MEG: Teaching that initial seminar at SLC and the work I have done since then in animal studies has certainly informed my research and absolutely shaped my dietary practices. This work got me to think about food in an entirely different way. But there are moments that can also push me in a different direction. For example, I just had a really hard conversation about Native sovereignty, animal rights, and food politics where there was a striking blindness to the everyday reproduction of settler-colonialism, and the ways certain positions and approaches can in fact work against Native sovereignty. And when I start to feel this paternalism or hear people say, "Native people should do X, Y, or Z," I question the politics of veganism. It makes me question my own dietary choices. And then I have to take a step back and talk to friends and colleagues and my partner to think about my position, how I am situated, and why I make the decisions and choices I make. I try to remind myself that these choices are messy and contextualized. I'm noticing that these kinds of conversations and some of my commitments around Indigenous sovereignty have begun to reshape my thinking about what I eat or don't eat as a particular kind of politics.

SK: I can see that. I've been thinking about how Indigenous people who are vegans might be mobilized in a problematic, tokenizing way. One of our other interviewees, Lauren Corman, also raises this question.

MEG: Absolutely. This happens quite a bit, the tokenizing and mobilization of those proclaiming Indigenous veganism, often by non-Natives, to impose particular viewpoints on tribal governments, for example. And this in fact takes away from a fuller and more serious engagement with Native epistemologies. Indigenous veganism offers a really wonderful opening to think more about the complexity of Indigenous sovereignty, knowledges, and food. But I am frustrated by those using this as a way to push a particular political project, in a way that disavows another political project. Also, it's not just about animals. For example, in Seattle we belonged to a CSA [Community Supported Agriculture] and we were able to get organic local products. But a lot of the organic berries are picked by Indigenous migrant laborers from Mexico in horrendous conditions, and this is one reason why we no longer belong to that particular CSA. So how do we think about justice in a way that is more encompassing of all of these complex political issues? It gets really hard, but I think it's important for us to think about. Especially with my son, I want him to be an ethical eater, but we are constantly developing what that means for us. I just want him to be thoughtful about how our choices are very much connected to broader structural systems.

Three years ago I became chair of an interdisciplinary program at UW called the Comparative History of Ideas, and we have all sorts of events—graduation, thesis presentations, etc.—that involve decisions about what food we are going to serve, how we will serve it, and whom we should buy from. I have been navigating these choices and trying to push an ethical approach as much as I can, but all the while moving slowly enough so that others don't feel that I am imposing my choices or ideas on them. So, depending on the different places we inhabit, I think we need to find ways of pointing out these kinds of entanglements and connections in ways that open up conversations rather than shut them down. That's one of the things I love about Jonathan Safran Foer's book *Eating Animals*; I was really struck by his ability to open up a conversation. This is crucial if we are ever going to continue to move together, however imperfect the ways in which we all move.

Another thing about that book—he begins by discussing his struggles with vegetarianism, and how having a child made it possible for him to commit to a vegetarian lifestyle. He writes that after the birth of his son, he received a note from a friend that said, "Now anything is possible again." And it's so moving and powerful. But when I had my son I struggled with that, because I had the entirely opposite feeling. I worried that if I raised him eating a vegan diet, he would not be able to go to Peru and eat my grandmother's cooking and our traditional food with my family. That was and still is a huge struggle for me. And then the first time we visited my parents in Virginia once he was old enough to eat solid food, my mother gave him some kind of chicken sandwich, which he devoured. I realized that in this context, maybe it was okay for him to eat different things. For my mother, feeding him is very much caught up in broader concerns about food as tradition and food as history. But this was interesting to me; I felt this kind of ethical responsibility to raise him in a particular way but then, as much as I am an anthropologist who writes against the notion of essentialism, I still have this fear of loss. What is he going to lose? It's tricky, and I think for me that's also why emotion is so important; these ideas come from deep thinking but maybe they come from some very deep feeling as well. Which is a different form of intelligence, a different form of engaging with the world, with ideas.

SK: And so when your son visits Peru, does he eat guinea pig?

MEG: No. During one trip to Lima when he was about a year and a half, my grandmother offered him ham and chicken, which he ate. This is when conversations with my partner began in earnest about what our son would eat or not eat, and they were tied to the significance of his knowing not just traditional *Peruvian* dishes but traditional and important dishes within our family. After that visit we spent a summer in Lima with him, but this was different because for the first time we actually had our own apartment. That was much easier because it was our space, which made me think we could possibly go to Peru for a year, and it wouldn't necessarily mean giving up on our commitments; it would just mean re-framing them.

SK: You partly addressed this already, but can you tell us a story about a time when your dietary practices have been the subject of awkwardness, celebration, or hostility?

MEG: Yes. My mother is this ball of contradictions; she loves Bill O'Reilly and she's a member of PETA. She's just a super interesting and difficult woman, whom I love to death, of course. One year for Christmas I decided that instead of buying people gifts we would sponsor an animal for each member of my family through the Farm Sanctuary. For my mother, we sponsored a turkey. And you know, you get a picture of the animal with a list of their food preferences among other things. Everyone seemed happy about this gift, which arrived in time for Thanksgiving. When we moved to the U.S., we didn't celebrate Thanksgiving; my mom would make a big Peruvian, Venezuelan, or Mexican meal instead of the traditional Thanksgiving meal. But my brother got here when he was six years old, so he's grown up in the U.S. and he likes the turkey and all the traditions—so over time my mom began shifting toward a more traditional version of this holiday. This particular year we went to visit them for Thanksgiving and she had made an elaborate meal that included turkey. As she is setting the table I see her get the picture that Farm Sanctuary had sent her. I remember the turkey's name was *Chiqui* and she put it right next to the turkey on the table. She thought it was the funniest thing. For me, it's just an example of the complete disconnect in our thinking about animals. She obviously understood what we were trying to say, but I think for her it was a deliberate way of minimizing the importance of what we wanted to do and making sure that it was still okay for some people at the table to eat meat.

SK: That must have been a tricky situation! Can you talk us through your typical approach to buying or otherwise procuring food?

MEG: We used to belong to a CSA, so every two weeks we would get a big box of vegetables and fruit from local farms. I would take a look at the box, figure out what I could make over the next weeks, create a menu, and then supplement that by going to Whole Foods or a local co-op. We try to be thoughtful about our choices. For example, in learning from Native leaders and activists about the conditions in which Indigenous migrants work in some of the organic berry fields in Skagit Valley, we changed our approach and no longer belong to the CSA that sourced from those farms. We just try to think carefully about the politics of the choices that we make and understand that we're not perfect, and do the best we can to minimize suffering.

SK: I think that's really important. Amy Breeze Harper recently organized a conference, the Vegan Praxis of Black Lives Matter, and one of the things that came up over and over, particularly from black activists who are also vegans, was frustration with white vegans who do not take racialized labor practices seriously. And then, on the other hand, you have these moralistic discourses about food, obesity, and related things.

MEG: Yes, I have similar frustrations. This is where thinking about Native sovereignty comes in for me. It's very much related to those concerns about imposing a particular way of thinking about the world, or food, without a full understanding. For example, the Makah whale hunt: As much as I might disagree with some practices, we need to understand the multiple ways in which this hunt—and other forms of killing such as salmon fishing—relates to Native sovereignty.

This relates to another point about the politics of killing, and care. I've been struggling recently with the idea of "killing *as* care." There's a famous case of a Bolivian woman who would rather kill her llama and sell her as meat to foreigners than sell her alive. So what does that mean in terms of thinking about those relationships, in terms of "killing as care"? Is it possible? In that case, the decision was linked to spiritual ideas about what happens when you sell an animal alive without knowing what the consequences are, without knowing what will happen to this animal. I find it is important to think in a multilayered way about killing rather than say, "Killing is bad." As much as I might have a hard time personally with some of these decisions, I've been encountering a lot of situations where people are pushing me to really think about this. So I try to be open.

CHAPTER 9

Disability and Interdependence

Sunaura Taylor

Even as a young child, Sunaura Taylor, now an artist, activist, and disability and animal studies scholar, understood that humans, animals, and the environment are intensely interconnected. Taylor's ecological orientation is not simply an intellectual focus but rather a set of political beliefs she endeavors to embody in her everyday life, though she admits that doing so is rarely easy. Taylor's work demands that audiences rethink the worthiness of vulnerability, of dependency, and of interdependency, particularly as these concepts speak to shared experiences among all living organisms in times of environmental turmoil and fragility. We discuss humane meat, vegan parenting, and the behavioral shifts that can occur when emotional knowledges synchronize with cognitive truths. We interviewed Taylor by Skype on February 16, 2017.

SUNAURA TAYLOR: Could you tell me a bit about the project, just to refresh?

SAMANTHA KING: Sure. The project emerged from several different directions, but in part it was inspired by an interview with Donna

Haraway, where she describes her complex relationship to eating meat. We were intrigued by her response and began thinking about other scholars who primarily write about multispecies relationships but who have not explicitly addressed questions of food in their work. We felt there was a gap in that literature. But we also ended up speaking with a small number of people like yourself, who *have* written about or addressed those questions. And we wanted to give scholars who address eating animals in concert with questions of colonialism, race, disability, sexuality, and so on a chance to talk about those connections and to address the normativity of multispecies ethics.

ST: That's exciting! It is fascinating to me how the ways in which we consume animals on a daily basis—whether through clothing or through food or whatever—is often absent within animal studies. So I love that you all are asking people to grapple with it more. I think that's great.

SK: Well, thank you. At some point you will get to read each other's interviews. It would also be cool to bring the contributors together in order to have that conversation. So stay tuned!

VICTORIA MILLIOUS: So we'll start at the beginning. Could you talk about when and where you were born and raised and your formative, cultural, intellectual, political experiences?

ST: I was born in Tucson, Arizona, and, no doubt the most obvious formative experience is that I was born disabled. From a very young age I had an understanding that my disability was caused by water contamination in Tucson. Specifically, it was military pollution that affected largely Latinx communities on the Southside and also a portion of Tohono O'odham land. So from a really early age that knowledge, combined with the fact that my parents were and are very creative, unconventional, and progressive people, created an environment where thinking about political and social justice issues was just part of our daily lives. Also, my mom had been vegetarian for a large part of her life, so thanks to her, including animals in our broader understandings of justice always made sense to our family. I just had this sense as a child that my body was the way it was because of how poorly people treat each other and the environment. That was the understanding I had as a young kid, which sort of gave me this identity as an activist. I felt this responsibility as a kid and I think on one level it also gave me meaning, gave me a way to understand being disabled.

The other really formative thing is that my parents unschooled my siblings and me, which is basically a radical form of homeschooling. At this point I was probably six or seven, and we had moved from Arizona to Athens, Georgia. My mom and dad really felt that a lot of the educational options available to kids were stifling to a kid's natural curiosity. So they unschooled us, which is the philosophy that kids have an innate desire to learn and that if you give them a supportive and enriching environment, they will go out and learn things, because that's what kids do—they're curious. That's how my parents raised us. I have three siblings: my older sister, Astra, who's two years older; my brother, Alex, who is two years younger; and my sister Nye, who is ten years younger. They were also, of course, extremely formative. Astra started a magazine, *Kids for Animal Rights and the Environment*, when she was about ten. Which was amazing! She edited it and sold it at the local health food store and through other venues. All the siblings and all of our homeschooling friends, we all wrote articles for it. So we definitely fed into each other's political thinking too.

SK: As someone who lived in Tucson briefly, I'm having one of those "How did I not know this?" moments about military pollution and the Tohono O'odham land. Were the pollution and its effects on people recognized publicly and addressed politically?

ST: Yeah, it's surprisingly not that well known outside Tucson, despite its being an important early environmental justice case. Affected residents eventually brought a major lawsuit against the responsible parties, and at the time the out-of-court settlement was the largest for any such case. Basically, from about the 1950s until the 1980s, Hughes Missile Company [now Raytheon] and other defense and electronics industries were burying their toxic waste in unlined pits in the ground. And they knew at a certain point that it was seeping into the aquifer and thus the drinking water, but neither they nor the city of Tucson informed the communities. In fact, at least some city representatives denied the problem by suggesting terribly racist reasons for why people were getting sick—suggesting they were just eating too many beans, for instance. Just horrible stuff. I'm currently researching this site for my dissertation, actually.

VM: That's really interesting, thank you. And we both look forward to reading your dissertation.

ST: I look forward to knowing exactly what the hell my dissertation is going to be!

VM: I can so relate to that! It took me many years. You've already addressed this a bit, but if you could move us a little forward in time, how did you become an artist, writer, and activist?

ST: My mom and dad are both artists: My dad is a scientist, but also a composer and musician. And my mom is a writer and painter, and she currently runs a small animal sanctuary called Dharma Farm. So art was something that was always in my home. I started drawing when I was super young, I don't even know when; I was always interested in drawing. And then when I was around eleven, we all got my mom a bunch of paints for her birthday and she started painting a lot, after a few years of not doing so much just because of how busy she was raising us. And of course, we were nerdy unschoolers, so we all wanted to hang out in the same room together and paint too! I just fell in love with it, and because I was an unschooler I was able to just spend all my time doing that. It wasn't like I had to go to school and then come back and I could draw for an hour; I could literally spend six to eight hours in the studio just messing around. That's how the painting thing emerged for me, I really just fell in love with it. Eventually I got an MFA at Berkeley in their department of Art Practice—which was such a great program. I graduated from there in 2008. While there I really started getting obsessed with veganism and animal issues. I had been a vegetarian since I was six, but this was . . . there is a cliché when people, especially vegetarians, suddenly realize that they really should be vegan, they just become obsessed, which was totally true for me. It completely consumed my life, to such a degree that I was basically like, "I have to make art about this! I have to write about this! I have to research it! I have to learn everything I can about it!" So I started doing all that, and that eventually led to my making these connections with disability studies—which I had been studying and writing about for many years already—and then writing my book, and then getting a PhD. It was this moment of being so overwhelmed by what is happening to animals on a daily level, and that I was contributing to that—that it really transformed my life trajectory, from being really committed to painting and drawing, to being more of an interdisciplinary artist/scholar.

VM: The paintings featured in *Beasts of Burden*, and other places, were those done during your MFA or afterward?

ST: Yeah, so I think the paintings of animals in factory farms that I did at Berkeley are largely to blame for this obsession that took over my life. Those paintings and my siblings, actually. I guess first, my siblings and I all became vegetarian around the time that I was six—basically we came across some sort of table for animal rights, and we all had this epiphany that meat was animals. It was one of those life-altering kid things where it was like "What! My world is turned upside-down!" So then we basically just said, "We're not going to eat meat anymore, Mom and Dad. You guys have to deal with it." My mom was into it because she had been vegetarian on and off for a long time, and my dad soon supported it too. They've both been vegan for years now. My siblings each became vegan when they were kids. But I didn't. And we constantly had arguments about it, and for some reason it just wasn't getting into my head. I understood rationally, but it didn't really hit me. And then, skip forward many years and I'm in this art program at Berkeley, and I had decided before I left that I really wanted to paint a chicken truck. Where I grew up in Georgia, you would see them all the time because there are so many chicken factories there. My siblings and I had this tradition that whenever we were driving in the car, we would hold our breath when we saw one. I think it started because it's such a horrible smell: the 100-degree heat of a Georgia summer combined with dying birds. But then it morphed into this morbid way of experiencing the intensity and awfulness of what was happening beside us. So I had this sort of emotional attachment to these chicken trucks and decided that I wanted to paint one. I ended up doing a painting that was about eight feet by ten-and-a-half feet. It took me a year, so it was basically what I did while I was at my MFA, and . . . if you spend a year looking at animals you learn a lot. I learned they were egg-laying hens, I learned about the industry, I learned that those hens, like all chickens used in industrial animal facilities, were disabled. And that sort of meditation that goes into focusing on something like that and doing the research for it was completely . . . I was just like, "Oh my God, I can't eat eggs anymore! My siblings were right all along!" I always joke that I had to write a book about this stuff because I had such extreme guilt that I was the last sibling to get it!

VM: Siblings have a way of calling some truths.
ST: Yes, they really do.

SK: Jonathan Safran Foer talks about that in *Eating Animals*. He says that we all know at some level that factory farming is horrific and that it doesn't somehow stop at meat, that it extends to dairy and eggs and all of that. But there's something about being ready to hear or feel that horror fully; it's not just about getting the logic of the argument but being open to it and affected by it.

ST: Yeah. And I had wanted to be vegan for years. I think it was my New Year's resolution probably five years in a row, to be vegan. But somehow I just hadn't gotten to that place where it became so obvious to me on a visceral level that it was easy for me to do—which it was at a certain point. I think it was also doing my own research. It's one thing to have other people tell you logical things, but I remember, for example, I was convinced that surely in the Bay Area I could find humane cheese and humane eggs. I was just like, "Okay, if they exist, I'm going to be able to find them here." Because I had always had this idea that "Well, but maybe this cheese doesn't come from a factory farm. You don't have to kill the animal to get this." There were all these little loopholes or something in my mind. Then once I started doing my own research I was forced to acknowledge that "No. Actually." Any way you cut it, that is a fantasy land. There simply are not female animals happily lining up to offer up their reproductive labor to human beings, which of course is what eggs and milk are. Humans always coerce it from them. But I had to come to those conclusions on my own. And for me, it was also realizing the connections between animal oppression and disability oppression that really forced me to start thinking about veganism differently. I understood disability oppression on such an emotional and structural level that once I was able to make those connections, then I could no longer *not* make them.

VM: So what do you see as the primary purpose of your work? Here we mean your artistic work, activist work, academic work.

ST: I don't think I could really say what my primary purpose is. But I can say some things that are sort of constant themes or values in my work. I think broadly I want to open up alternative ways for people to think about and experience disability, as well as their understandings of and relationships to the nonhuman world—and I see these two processes as entangled. There are also certain concepts that are really central to me—vulnerability, interdependence, dependency—even as I'm still trying to figure out what they mean exactly. These concepts, which are so central to disability studies, seem especially important to bring forward right now in terms of climate change and mass extinction, and so part of my work is to expose these central disability themes as vital to thinking across species. But it doesn't just go in one direction, and so it is also a matter of showing the importance of thinking ecologically and across species to creating new and generative ways of understanding disability. Ultimately, of course, I'd say that I'm interested in imagining what liberation across species could look like, and I hope my work is working toward that.

SK: I think you probably already answered our question about how you came to be interested in animals and animal ethics, but is there anything you want to add?

ST: I would just emphasize that it was so many different factors that brought me to the point I'm at now. Meeting disabled people who were themselves vegan and animal advocates was also this significant moment . . . because it's so hard for me to cook; it's so hard for me to do that sort of work. So I always had an excuse that was also very valid: "Well, I already have such a hard time eating" and "I have other people help me cook and stuff, so this is going to be more work for them." That was also part of the narrative in my head. So meeting other disabled people who were making those choices despite similar challenges, I think was also a major motivation for me to become vegan.

SK: What kinds of things made that possible? Was it just a matter of asking the people who helped you to do different kinds of things? What did that look like in practice?

ST: In practice there were quite a few things. I talk about this in my book: that becoming vegan was perfectly aligned with coming to terms with the fact that I needed help, and that I needed to hire someone to help me. So in that way I felt that the coming into realization of veganism and coming into a political awareness of disability, and not feeling ashamed that I needed help, was actually about figuring out a practice of interdependence. I think there's a reason those things happened at the same time. So pretty soon after I became vegan I started hiring someone to cook for me twice a week. In California you can get funding for in-home care, which unfortunately in a lot of places you can't. I was also extremely lucky because I have a partner who became vegan very quickly; it was probably only a few months after me that he became vegan. And my family is vegan, so I was already in this very privileged position where it was pretty easy for me in terms of my social network to do that. I'm also not someone who has a difficult time digesting certain foods; I don't have a lot of allergies. I think a lot of disabled people often have more complications in terms of eating and in terms of who is helping them. Are they in institutions? Do they have any choice about who is helping them? All those factors are really important to consider, in terms of disabled people and their choice—and sometimes lack thereof—of being vegan.

SK: Could you tell us more about how and when you started to think about the connections between the oppression of animals and the oppression of disabled people? Perhaps you could also discuss your forthcoming book, and what animal studies and disability studies have to learn from one another.

ST: The first entryway into making these connections for me was actually with the painting I was doing of the hens on the chicken truck. The more research I did, the more I realized that animals in factory farms are disabled. They become disabled because it's such an extremely brutal environment, but they are also bred to be disabled—it is what makes them profitable. The more I read the more I realized that disabling animals is not incidental to animal industries. It is essential for the work they do and the profit they create. Virtually all animals used in food production are in fact manufactured to be disabled, with bodies that have been bred to produce so much product that the animals are impaired. A similar pattern of what I would call "profitable disablement" can be seen across a wide variety of animal industries, from fur farms, and animal research labs, to zoos and circuses. The very thing that makes an animal profitable is disabling.

Thinking about exploited animals as disabled led me to start considering how ableism as a system of oppression affects nonhumans. And in many ways that is actually really what my book is about. It is really a way of looking at ableism expansively, demonstrating that ableism oppresses everyone, including nonhuman animals. I do this by considering how intellectual and physical capacities are used to justify exclusion—right? Animals can't do this or that so we are justified in demeaning them—but also by delving into the concepts I mentioned above, such as dependency, which is fraught with negative connotations and is often associated with both disabled people and domesticated animals. Ableism forms our worlds in all sorts of different ways. But what I'm looking at particularly in this book is how ableism, at least within Western worldviews, affects the ways in which we understand

and think about animals as lacking, as physically and intellectually inferior in all these different ways.

And finally, the book also really tries to investigate histories of dehumanization and attempts to expose shared genealogies of disability and animal oppression. I try to grapple with histories of demeaning animal comparisons while also making room for the importance of acknowledging that we are all animals. I am interested in the spaces that can open up when we recognize that we are animals. I have a few moments throughout the book where I write about the way my body moves in ways that are very animalistic. I don't use my hands; I often use my mouth to do things. The ways in which I feel animal in my body and the ways in which that doesn't have to be a negative experience—that it in fact can be a moment of recognizing shared experiences with other species, and also shared vulnerabilities.

At the same time, I am careful to recognize that part of why I can go there is that I'm white, I'm not intellectually disabled, I don't have a communication impairment. Right? There are multiple ways in which identifying as or claiming animal is just far too dangerous for some people—people who historically have been animalized as a way of justifying their exploitation or murder. I am very careful in the book to recognize that as much as I want us to think about how we are all animals, and the radical importance of doing so, I also want us to simultaneously hold onto the fact, as philosopher Licia Carlson does, that for some people the most radical thing is to claim their humanity. Because that is something that has been denied to a lot of people.

SK: The talk you gave in Alberta has been so helpful for my thinking and teaching—to think about shared histories and how notions of dependency "travel." Have you had any pushback on these ideas? While you are careful to recognize histories of dehumanization and ongoing dehumanization, it's also important to you to acknowledge that we're all animals. How do people respond to those claims?

ST: My book hasn't come out yet, so I really can't say, but I'm assuming I might.

Animal rights and disability rights have very often been presented as at odds, or at least in conflict. There are many reasons for this: the troubling and ableist ways that disability has been used in some popular animal rights arguments—Peter Singer's work for example; the mainstream animal rights movement's obsession with health and physical fitness; and of course because of the ways disabled people have been dehumanized historically. My book is really trying to show how unfortunate, problematic, and unhelpful this framing is, but it's a pretty powerful one, and so I'm sure there will be people who push back against even the idea of linking animal and disability justice. Of course, my hope is that I am thoughtful and careful in my writing, but there's no doubt that bringing animal and disability studies together has risks. There's no doubt that asking people to be open to the fact that they are animals has risks. So I'm sure I will get pushback. But I don't think I have a sense of what that will be like, or even from what angle it'll be pushing. I may just get a lot of pushback for the fact that my book is very much arguing for veganism as an embodied political praxis. Who knows, I may get pushback that it isn't vegan enough! So I have no idea, but if I do I'm lucky! That means people are at least reading it, which is great!

VM: Yeah, conflict at least means people are paying attention to some degree, which is an opportunity.

SK: Well, we're excited to read your book.

VM: We know you've written about this in various articles, but can you talk about your views on humane meat?

ST: Oh yes! I got to the Bay Area right around when Michael Pollan's *Omnivore's Dilemma* came out, and I felt like there was this whole vegetarian backlash—a belief that vegans and vegetarians were naïve. For me as a vegetarian at the time, instead of pushing me to reconsider eating meat, I found the arguments for humane meat to be so troubling. Now I understand that I found them troubling because when you really get into the nitty-gritty of them they are quite ableist. Now I understand that, but at the time I just thought, "Why are these arguments for humane meat making me want to be a vegan?!" And sure enough, I would soon become one! In *Beasts of Burden* I spend quite a bit of time unearthing and analyzing the rhetoric around dependency that "conscientious omnivores" use to justify exploitation. In short, the argument is often that because domesticated animals have entered into a sort of evolutionary social contract with us, where they are ultimately dependent on us for their care, we are justified in using and slaughtering them. Disability studies has long examined how notions of dependency are used to excuse the oppression and marginalization of people, complicating our understandings of what dependency means and why it is so despised. So I use these tools to challenge the way arguments for humane meat often use rhetorics of care and interdependency. Ultimately, I argue that this whole framing effaces all sorts of power inequalities and oppressive structures that are—conveniently for humans—rendered as natural and outside of human social systems.

There are so many ways in which mainstream conversations around humane meat are just so misleading and frustrating—from an environmental perspective, it just allows people to ignore the unbelievably clear research which shows that animal farming is a major factor in driving climate change and in so many ways the current mass extinction crisis we are in. I will say, however, that thinking about humane meat through a lens of food sovereignty instead of through a liberal American "freedom to choose what kind of meat one has" perspective, does complicate these issues for me—not in a "Let's eat meat now" way, but in a way which demands that environmental and animal issues be understood as absolutely imbricated with histories of racism, settler colonialism, imperialism, and class dynamics. Obviously different movements for food justice are working toward different things, and the community that I'm mostly pushing back against in my work is a more mainstream one.

VM: I'd say that's how humane meat resonates in our locale, here in Kingston—also in terms of who can afford it, where is it available, and whom those types of stores cater to. We'd love to talk now about parenthood. Can you tell us about the relation between vegan politics and animal and disability questions, and your approach to feeding your children and parenthood more generally?

ST: I have a daughter who is almost two and I joke that she is going to write a book called *Vegan Since Conception*! I had a vegan pregnancy, which was great, and she's two, she's vegan, and she's doing great. On one level all of the medicalization that happens around being a very visibly disabled woman who was pregnant totally took away from any concern people had about my being vegan. It was like my veganism was so low down on the list of what people were freaked out about that it really did not come up. Whereas I've heard from so many other parents that their doctors or midwives really pushed them on their veganism.

I'm very grateful that my brother, Alex, has kids and they're quite a bit older than my daughter, so I'm already asking him, "How did you talk to your kids about veganism? When they're at a party and their friends are eating nonvegan cookies, how do you do deal with it?" I still don't really know how I'm going to deal with those things. I'm certainly never going to tell my daughter she has to be vegan. But I want her to be raised aware of disability oppression and of gender inequality, and of challenging white supremacy, so I feel like animal justice on one level is similar to those things. I'm vegan because I'm a feminist, because I'm anti-ableist, because I don't believe in capitalism—so my hope is that veganism won't be seen to her as being about food restrictions but will be understood as part of our family's beliefs about justice and the kind of world we want to help make. That somehow it's not just about food; it's about her having an awareness of these larger social justice issues so that then she can make her own decisions.

VM: How do you inhabit veganism in the home and in your daily life? What does that look like?

ST: One thing that became important to me as I was writing my book was thinking about veganism as an anti-ableist practice. I don't think it's helpful to frame veganism as a diet or lifestyle choice. It's very much a part of the way in which I want to embody my political beliefs on a daily level. But part of that is also acknowledging how hard it is. For example, I talk about my relationship with my dog Bailey a lot in my book, and I challenge a certain strain of animal advocacy that has argued for the eventual extinction of domesticated animals. I push back against that a lot because I feel those arguments are often rooted in ableism and ableist narratives of dependency, and what it means to need help. So I think a lot in the book about what it would mean to have relationships with animals that are really interdependent and supportive: He needs me to care for him; he also cares for me on emotional levels, so how do I make that as non-exploitative as possible? How do we take seriously the idea that we can listen to animals and try, as impossible as it may be sometimes, to consider what they need and how we can help provide that, while also always remembering how much animals offer and give to us? It's not a one-way street; we are totally entangled and interdependent—sometimes in really uncomfortable ways. But then at the same time I'm also a crip-mama to a toddler; I'm in a PhD program; I'm totally exhausted and so sometimes Bailey's walks are just way too short and I realize, "This is not okay." But at the same time, it's what is happening right now, so I don't know. I don't think it's something that I have remotely figured out how to do—how to really be in an ethical relationship to him; what does it actually mean to be interdependent with another species? And how do we care for animals that are reliant on us in ways that are as supportive and as non-exploitative as possible? It's a really significant challenge on a daily level. And these questions haven't really touched on other creatures that are less pleasant to us, right? But I think trying to consider what being in ethical relations across species means is an important thing to do.

VM: Can you tell us a story about a time when your dietary practices have been the subject of awkwardness, celebration, or hostility?

ST: It's weird because I feel like on some level being disabled, first, already alters the normative routine on such an extreme level. So again veganism is almost always this secondary thing, where if I go out for dinner, or to someone's house for dinner, people are far, far, far more concerned about the fact that they might have steps to their house, or am I going to be able to fit through the door, or just their own discomfort in dealing with the way that I eat, than they are with *what* I'm eating. So I think on one level, being disabled has shielded me a bit from a lot of the discomfort that many people feel—which isn't necessarily a good thing, right? This often comes from ableism. I think sometimes people have so much anxiety about my disability that they don't see my veganism, and other times people write it off, people are just like, "Oh, she's a disabled girl and isn't it sweet she loves animals!" There's something that can be pretty condescending about it too. I also should say that I am in a really unique position in terms of the community I have around me: I have a vegan family, a vegan partner, and many—though of course not all—vegan friends, which I know is not a common experience for a lot of vegans.

VM: Can you describe a key dilemma or question that haunts you?

ST: I would say it relates to practicing these ideas on a daily level. So often people critique vegans because there's this stereotype that we think we're being cruelty-free, or that it's this ideal that can't be reached. I think vegans need to take that back and say, "No, veganism is not about being cruelty-free or practicing ideal nonviolence." It's actually a way of saying that it's really incredibly hard to actually figure out how to live in ethical relationships, whether you're trying to do that across human difference or whether you are trying to make a space that's accessible and open to other sorts of species; the challenges are really real—but veganism is saying that it's absolutely essential that we try.

It's this difficult space of relating, of how to live *with*, that veganism embodies to me. I want my veganism to challenge a sort of "nonrelating" brand of animal rights activism that suggests humans just need to leave animals alone and not relate to them—because it's impossible and also it's not the world that I want to live in. But I also want my veganism to challenge a certain romanticization of human and animal relationships that can too easily ignore power inequalities and exploitation. I think Claire Jean Kim has a wonderful quote at the end of the Introduction to a special issue of *American Quarterly* that she and Carla Freccero edited. She says that it's not enough to love animals. I think she's referring to a sort of romanticization of having a relationship with an animal and loving that animal or loving the practices that are around those relationships with animals. I think what she's saying is that when you do love something, it's really important to also realize that love isn't enough. It's not enough for my dog Bailey that I love him; I also need to give him longer walks and think ethically about what would make his life better and try to really practice those things.

SK: It's been really incredible talking to you, and you've made me think about some of the assumptions about disability that are built into our questions, so that was a great learning experience for me. Is there anything you want to add?

ST: Not at the moment, but I'm sure I will think of things. This has been really lovely; I appreciate it. And I really look forward to the book because it's getting at some issues that are really hard to grapple with.

CHAPTER 10

Asking Hard Questions

Neel Ahuja

Interdisciplinary scholar Neel Ahuja discusses how his early animal rights activism has informed his pursuit of questions concerning the animal and the environment within postcolonial studies, and how postcolonial studies is critical to how he views the fields of posthumanism and animal studies. Naming veganism as a "limited tactic" that works to imagine ecological futures distinct from the period of mass extinction in which we live, Ahuja stresses the importance of conceptualizing "species" as a political question and not simply an ethical one. He offers sound advice for early-career scholars who must navigate eating and interviewing simultaneously, along with his thoughts on the potential for kinship and the politics of the killjoy. We interviewed Ahuja by Skype on September 21, 2016.

SAMANTHA KING: What is the primary focus of your academic work? Why you do what you do?

NEEL AHUJA: I think that is a question a lot of people are thinking about, regardless of what field or discipline they are working in these days. Especially with the privatization of university education, there's a whole discourse in the humanities now about the "university in crisis," particularly about the idea of critical thinking and the position of the critical intellectual as something being monetized. So is it possible to have the purpose of one's academic work exceed the purview of what the institution says? It is an interesting set of questions to think about. In the U.S., a lot of this discourse is being generated around how graduate students are being treated and how their prospects for employment are shrinking, how they're being turned into a permanent precarious class of employee. In that context, I actually think there are moments to resist some of the ideas about "rejecting the institution" as a whole and perhaps instead fighting for its invigoration. I think that critical academic work has an important place in that.

Coming from training in postcolonial studies and particularly in the literary–cultural aesthetic track of postcolonial studies, there is this inside/outside dimension to where the work fits into academia. There's a relationship with social movements, and much of the work is an attempt to reconstruct an account of colonial modernity, of how we come to the colonial present by looking at various histories, imaginaries, social practices—including eating and agricultural practices—that have been erased, marginalized, and suppressed by the current configurations of state and capital. As someone who works in postcolonial studies and who is trained in thinking about representation—about literary, cultural, aesthetic forms—I do feel that academic work has instrumental purposes. It is meant to engage with certain questions that exceed the bounds of the academy. In its inevitably constrained position inside educational institutions generated by colonialism, postcolonial studies attempts to extend intellectual projects that were generated by liberation movements coming from what we can now call the Global South against 500 years of colonial rule. Part of my academic work has been to grapple with how that extends into the realm of interspecies relations, the ways in which the planet, land, species—including humans but other species as well—have been "engineered" into colonial modernity and how we might reimagine worlds which have been suppressed by that colonial modernity, and worlds that might yet come from it—or through or against it. The academic work doesn't always mean putting all of your political ideals or agendas on the table immediately; it does involve some reflection, depending on what you're writing and what the object you are analyzing is. It may be relatively technical or it might not bring the biographical in at every moment. But I think for me the university is an institution where this kind of imagining can or at least should still take place. It's unfortunate that to a large extent, the left in North America and Europe has increasingly been forced to retrench into the university, *out* of the public sphere in some sense. But that's something I hope we can continue to struggle against. So I think regardless of what fields or disciplines we're working in, the move to corporatize the university and turn it into a career-training field for students is something that works against these broader intellectual currents that at least I'm invested in.

SK: That's a great answer. I appreciate how you took a broad structural approach and discussed the corporatization of the university and academic precariat. Bringing in those issues adds a dimension to the conversation that hasn't arisen in other interviews. It also makes a nice segue into the next question, which is how you came to write about what you call in your book the "government of species." Maybe you could also talk about your key intellectual influences as you did that work.

NA: Of course. You mentioned Donna in that earlier discussion we had. Haraway's work has been very influential for me, although I will give you a different answer to the eating question than she did. Her work was, for me, influential in terms of the questions and options I thought about as a grad student, but it also offered a pathway to thinking about relations that seemed to be divided; human–animal or nature–society are common divisions that often ground the disciplines that we work in—certainly literature, when you're trained in a literature department like I was where, let's say, in my case, the Frankfurt school was a critical set of conversations that grounded our training. And that Marxist approach to understanding and reflecting on aesthetic questions tended to question in what ways capitalism and other structural forces determine the formal qualities of our aesthetic lives, including literature and other cultural products. And Donna's work was really inspiring in that context of criticism, especially her early work which is really explicit about its Marxist–feminist commitments. Her work is great about toggling between the aesthetic and the material and understanding how the aesthetic can mask the material, but also work to reform the material. So, as opposed to some other major names you might associate with animal studies more recently, Donna's work stood out to me before animal studies was turned into this field that could become institutionalized.

That said, in the introduction to the preface of the book I'm also trying to think between feminist science studies and postcolonial studies. In terms of how postcolonial studies informs my work, it is critical to how I end up viewing a field like animal studies or methodological orientations like posthumanism or new materialism, for example. The question of nature is something that has been repeatedly invoked throughout decolonial struggles prior to the institutionalization of postcolonial studies as a field, but it has also been really significant to how postcolonial theorists conceptualize transnational power formations. We'll get into this with some of the other questions, but thinking about the "government of species," that concept I put forward in the book, comes about, for me, by thinking about colonial sovereignty as a problematic that is challenged and theorized intellectually through decolonial struggles and then later through second-order reflections on those struggles in a field called postcolonial studies. Part of why we need second-order reflection on them is because postcolonial studies is all about the limits of anti-colonial nationalism. And we might consider questions about animals and the environment as being one arena in which anti-colonial nationalism has failed to substantively redress legacies of colonialism. It was important for me to put forward a theoretical device for integrating these two perspectives, postcolonial studies and feminist science studies, which is how I began to think about the concept of "government of species." Thinking about *species* as a term for politics, rather than a biological term or a term for ethics, was really important. Part of the theme of this, for me, is that the animal question is a political question, not simply an ethical question.

SK: Could you say more about that? This has come up in earlier interviews. Why is it important for you to approach the animal question as a political question, and what gets lost when we approach it as an ethical question?

NA: There are two ways I would like to address that. The first one is more biographical, and the second one involves thinking about these fields and how they're institutionalized. So I have been vegan for eighteen years, and part of my political education involved my coming to understand the ways that industrialized factory farming works in this country, and joining organizations that were working to contest the ways that industrialized factory farming has negative, deep violent impacts for many species, including humans, and for the broader planetary systems in which we live. A significant part of this political education for me was my coming to an ethical understanding about the recognition of the lifeworlds and interests of nonhuman species. I still believe that this is too often and too quickly dismissed in some sections of the left. As a teenager, I saw one of these flyers in a bathroom stall that had pictures of what goes on in a slaughter operation and I decided at that point that I was going to look more into what industrialized factory farming was. That said, I also have this family history of coming from northern India; my parents immigrated to the United States in 1965. Prior to 1965, there was very little Indian immigration to the United States because the quota system gave priority to European immigrants, particularly western and northern Europeans. But in 1965, the first year Indian immigrants were given broader quotas for entry, my parents came to the U.S. And we have a tradition of vegetarianism in my family, one that is not necessarily oriented—at least in my family—around the ethical question of what it means to eat animals or to eat their bodies and body products but is more an offshoot of a Hindu nationalism that was deeply entrenched in right-wing communal struggles that were targeting Muslims and Christians. This isn't something I necessarily understood growing up. Being exposed to arguments about the ethical protection of animals was important, but I couldn't understand my family's own dietary practices in that light necessarily. In fact when I first became vegetarian as an eleven-year-old, and then vegan as a seventeen-year-old, my parents were resistant in both instances, despite the fact that we largely ate vegan anyway. So the way that animal activists' writings often focus on the ethical dimension of eating animals in some sense feels somewhat provincial to me, knowing more about this history now. And also knowing about the global inequalities wherein meat eating is experienced as a luxury in most parts of the world. Thus to treat a prohibition on eating animals, or animal products, as a universal ethical ideal is a little bit troubled by the fact that many people, for the majority of their diet, are default vegetarian; they may want to eat meat, or they may occasionally be able to eat meat. To view meat eating as simply an ethical transgression, I think, misses the political context behind these choices. It suggests there is a liberal individual who operates freely in the world and is unconstrained by the forces that, in some sense, determine our diets. There are countries in the world like the United States where huge agricultural subsidies support the animal agriculture industries, making animal-based foods artificially cheap. The capacity to be vegan grows with one's economic means in such a society. I don't think that is *always* the case; there are people who engage in vegetarianism and veganism across class boundaries. I am acutely aware of the ways that veganism can be mobilized as a progressive class politics, and also a regressive class, communal, or religious politics, as it has been in India. That is one side of why I think about these questions around eating from a political rather than an ethical perspective. I think of veganism as an important yet limited tactic for people in countries where there is an advanced factory-farming system that is industrialized.

The second part to this is that postcolonial studies comes out of a moment in India, involving the Subaltern Studies Collective, a group of historians who are remembered as formative for the development of postcolonial studies as a field. They were interested in agrarian questions, and their work, particularly that of Ranajit Guha, focuses on peasant insurgency in rural areas in India. Postcolonial studies as a field comes about in part because of crises of nationalism that occurred after the independence of India, one of which has to do with the fact that people who live in rural areas, the peasantry, the Dalits, and the Adivasis in India continue to face horrible discrimination. In many cases they're still living off the land or in pastoral forms of agriculture. I don't think the rise of animal studies or food studies in the Global North has adequately grappled with the agrarian-studies basis of our anti-colonial critique. That is something I have a ton to say about, but I am going to cut it off there. These anti-colonial struggles are always struggles about nature and food, and I think that although agrarian studies is a kind of unsexy field compared to animal studies or food studies, it is really important not to dismiss that area.

SK: Thank you for that complex and provocative answer. When you said in our e-mail exchange that you were interested in exploring food studies and animal studies from the vantage point of colonial geopolitics, anti-colonial nationalism, and postcolonial studies, is part of that project to reconstruct a different genealogy of food and animal studies?

NA: Yes. That is something that, especially earlier in my career, I was already interested in. The first essay I published in animal studies was a short piece called *Postcolonial Critique in a Multispecies World.* That was a moment when I was reflecting on the extent to which animal studies was starting to become institutionalized in North America, Britain, and Australia—and missing these other histories of inquiry about human–animal relations. The fact that none of the main animal theorists have grappled with Gandhi's body of writing is a mystery. Not that I am really into Gandhi; I actually find his politics to be terribly regressive. But I think his animal writings are super interesting, and he really developed an ethical discourse about animals, one that may not necessarily be legible as a kind of animal rights or animal liberationist view on human–animal relations—in part because Gandhi was grappling with how to feed a largely poor country that was coming to independence. How do we grapple with the logistical challenges that arise when you're building a national agriculture? Keep in mind that under colonial rule famines were a regular occurrence in India. Even though hunger is still a huge problem there with rising economic inequality, the life expectancy in India shot up—way up—after independence. So there are reasons Gandhi didn't turn his own ethical practice of refusing animal products into a universal ethics for all Indians. There are moments in his writings on animals where he says, "You must demand of the animal the same respect that you accord it." I don't know how that would work exactly, but he claims that if in India monkeys are considered agricultural pests, and if the monkey cannot respect your own needs to cultivate the land, you may need to eradicate the pest. He writes this even though his own dietary practice was in most cases vegan: He hated milk, and in India eggs are considered a nonvegetarian food. Gandhi went through periods where he just ate fruit and went through a lot of questionable dietary experiences; we don't need to get into all of that. But this is one symptom of thinking of the genealogy of animal theory beginning with Peter Singer. By that point you've already reached the neoliberal moment, the moment where the West is de-industrializing and where it can't conceptualize food practices that are widely decentralized, where peasant and pastoral agricultural practices are the norm. That's part of why I think questioning the very field formation of animal studies and food studies through agrarian studies can be a really useful exercise in trying to think historically about how we've come to define our analytic objects in these ways.

SK: We may have touched on this already, but I wonder if you want to say anything more either about the critique you offer in the preface of your book regarding the racialized or orientalist tendencies of posthumanism, or about these bigger questions around colonial geopolitics in relation to food studies and animal studies.

NA: The only thing I would add about the book is that the preface is roughly oriented toward thinking about the theoretical turns in the humanities and social sciences toward posthumanism, animal studies, object-oriented ontologies, and new materialisms—where nonhuman objects play significant aesthetic, political, and social roles and might be considered actants, with their own sites of agency. So for me, I do think it is an arbitrary exclusion to say that animals and other species have no political agency, or produce no historical change within the political domain. But I don't think it's particularly useful to do this analogical operation where we say, "There are established political struggles around race, gender, sexuality, nation, and 'species' is one of the next frontiers of it." Martha Nussbaum calls species one of the frontiers of justice. I think that analogical form of thinking is also at the basis for a certain tendency in animal studies to conceive of human violence against animals as being this broad scope of violence from which we can say, "Look, human violence against humans—this is a form of animal violence that is a small portion of the larger, overwhelming interspecies violence that exists in the world." There ends up being a weighing of oppressions that I think is completely counterproductive for producing actual political solidarities around interspecies questions. I think it's also totally ahistorical. If you think about colonialism as a process that has worked in tandem with the rise of capitalism, part of what colonialism is about is appropriating what Jason Moore calls "cheap natures": the labor and energy forces of species and environmental systems that can be used for accumulation once those forces and labors are transformed into commodities. So specific historical conjunctures come about in which both human and nonhuman energies are increasingly turned into labor for the production of certain forms of capital, and for reproducing types of circulation and types of national state formations that increasingly appropriate human and animal bodies. I don't think there's been this primeval, you know, "Humans have always exploited animals" and "Humans have always exploited each other," and one of the two is bigger than the other. There are systemic forms that bring different bodies into relation through colonialism and capitalism, and it's necessary to think about the government of species in terms of these systems rather than having an objectified field of identities that can be crossed with one another.

VICTORIA MILLIOUS: We can always come back to these points, but we'll move along now to some of the more personal questions. You've touched on this, but can you describe your relationship to consuming meat or other animal products, and to what extent this has been an issue or preoccupation for you? How have your ideas and bodily practices with regard to animal consumption evolved?

NA: Sure. As I mentioned, I have a tradition of vegetarianism within my own family. It's not that everybody was vegetarian, but that certain relatives who were active in specific organizations associated with the Indian nationalist movement gave up meat. Also the general practice was that women did not eat meat, up until the last generation. Now my relatives in India basically all eat meat. This was interesting because my mom doesn't actually eat meat—I have never witnessed it at least—but chose to raise me as a meat eater, including feeding me beef—which would *not* fly among my family in India. The current government in India has been encouraging a kind of cow protection vigilantism and beef is either explicitly or implicitly banned in most parts of the country, beef in particular. But I grew up eating this meat. And at age eleven, after a debate we had about vivisection in my fifth-grade class, I decided I wasn't going to eat meat anymore. I mentioned the bathroom flyer I saw at age seventeen, and within a few months I decided to give up all animal products. When I went to college, we started a campus animal activist organization and for three years a close friend and I ran that organization. So I was very much involved in the mainstream animal rights movement for a few years. During that time I met a lot of national activists and was lucky to be exposed to an organization called Feminists for Animal Rights. Although their own writings weren't necessarily focused on race and colonialism, they helped to gather activists who were thinking intersectionally about species, race, colonialism, gender, sexuality—outside some of the main animal rights conferences. This was really influential for me around age twenty when I was going to these meetings and realizing that the mainstream animal rights movement was very particular about excluding these other forms of inquiry.

Around that time, I also worked as an intern for one summer with the Physicians Committee for Responsible Medicine [PCRM] in D.C. I was researching U.S. agricultural policy for them, so I learned the ins and outs of the farm subsidy system, and how that system was largely oriented toward building cheap meat, and how it covered that by making grain subsidies the primary target. So monoculture grains are farmed in huge amounts as animal feed, but if you focus on producing large amounts of monoculture grains, they can theoretically also be used for human consumption. They can be considered a food security thing, and it doesn't look like you're promoting very narrow interests of specific commodity agriculture. This is why meat subsidies are basically disguised as food security for the U.S., but so much of that grain is used in an inefficient system to feed animals rather than feed humans, and then there is a much smaller amount of food based on all that energy input that comes out as large slaughtered animals.

VM: What year was that, when you did this internship and research?

NA: It was in 2000. These were the early years of my work as an animal activist and PCRM was a great environment to be in. Because although they have some connections to PETA, which I always hated, they really had an interesting set of agendas that were focused around questions of health policy and diet, so that was a space where I was able to explore how animal questions are related to questions of human inequalities and violences. And even though it was focused on the national context of the U.S., later on I was able to think more broadly about how that connected to other countries like India, and many countries around the world that had been ravaged by the "Washington Consensus" neoliberal attacks on their trade barriers and their agricultural subsidies. All of this is something I was starting to think about at the same time, and by the time I was a junior and a senior in college, I ended up moving on to starting other types of activist organizations. One was a labor rights organization; it was focused on international labor questions and neoliberalism. The critique of neoliberalism became really significant there; to me it seems to be very related to what was going on with agriculture subsidies, what was going on with the transnationalization of manufacturing labor that we were focusing on. Those were some beginnings, and then in graduate school I had the opportunity to read broadly in anti-colonial theory and postcolonial studies; it was there that I really built a global perspective on these questions. But for me they were always questions about political organizing, as well as being broad ethical questions. These were always questions that came up in the process of political organizing for various liberation movements.

VM: Can you discuss if and how your writing about animals and other life forms has shaped your dietary practices or vice versa, or, more generally, the relationship—if any—between writing and eating?

NA: Most of my writing on species concerns primates and vivisection, and I'm only more recently writing more about eating, and animals as food—although it has come up in some writings so far. I do think that in the whole history of my activism, thinking about food was an important entry point into broader political struggles for me. I'm not sure that starting to write about animals has really fed into my dietary practices so much—it's remained pretty constant since I was a teenager. I guess I'm not really sure about that. But I certainly think that thinking critically about human–animal relations, and my work in the animal rights movement, and my becoming a vegan, certainly had something to do with why I pursued how postcolonial studies could think further about the question of animals, the question of environment. But I'm not sure that the actual act and practice of writing has shaped my eating, necessarily.

VM: That makes sense. We haven't interviewed many longtime vegans. We've heard very different answers to these questions, and even within the group of researchers writing this book, we're still grappling with our eating practices in some ways. It's unique to hear a response from someone who's had a relatively consistent approach to their own diet.

NA: Yes, I think the only struggle I have with eating now is that since I've become an academic, everybody seems to want to talk about wine and stuff.

SK: You find people talk too much about wine?

NA: Yes!

SK: Right, and the bourgeois preoccupation with what we're putting in our body, and where we're sourcing it, and all the localism stuff. . . . I've noticed those topics take up more and more conversational space.

NA: As someone who's really committed to working in public institutions, I really think it's weird that it's kind of standard practice for academics to spend huge amounts of money on fancy meals when doing job interviews. It's very strange to me. I really think that public funds should be spent in a responsible manner.

SK: We do some of that in Canada, but we're not usually permitted to spend money on alcohol.

NA: Right. And in California, that was also really different. I still am living in North Carolina but I'm about to start at Santa Cruz, and I don't think they allow alcohol to be purchased during these types of events.

VM: Can you tell us a story about a time when your dietary practices have been a subject of awkwardness, celebration, or hostility? I'm particularly interested in what it was like for you on the job market, doing campus visits. Has anything memorable happened? Did you feel you were performing your own veganism, or has it been a subject of interest or curiosity?

NA: Yes. Before I give a talk, or do a campus visit, interview, or any kind of invited event, I always bring up the fact that I'm vegan. I'm not someone who otherwise cares too much about what food I eat, so I'll just eat anything available that's vegan. But in some instances your hosts will take you to places with limited options. This was probably the case when I came to interview here at Chapel Hill in North Carolina. I remember when we were eating at restaurants during the interviews, there wasn't much that I could order. But I expected that going in. . . . Anyway, advice to grad students who are vegan and interviewing: Just eat before you go out. Both because you may not have anything to eat when you go out for your interview meals, but also because those are really *interviews.* Those aren't really meals, and you need to focus on the conversation. I found it was advantageous to already be full when walking in to the lunch, where eight people were interviewing me. Those weren't super awkward for me.

What might be a more useful story is from when I was in college. I lived in an intentional community: a housing cooperative with an explicit social justice and environmental justice mission. And for three years I was the food manager there, as one of the vegans who lived in the house. There were other vegans, but it was a vegetarian cooperative, mostly vegan, and I coordinated all our shopping and organized all the meal planning to ensure it was primarily vegan. When people who were cooking for the co-op wanted me to buy an egg ingredient and I basically told them "No," those were definitely awkward experiences. I think this is a really interesting difference between the United States and India to think about: In India egg is often considered a nonvegetarian food, and in the U.S. it's considered vegetarian. This was a subject of total confusion repeatedly in our discussions. For me, it had something to do with knowing about the specific factory-farming practices around chickens and eggs, but I think because of the way people draw up these categories of what's "meat" and what's not meat—why isn't milk meat? Does it have something to do with the solid form of the object as opposed to its being a liquid? What's the difference between different types of animal products and how you categorize them? That can lead to all kinds of confusion and awkward experiences. I have friends who are vegetarian and vegan who work in China, and they talked about how, yeah, you can get vegetarian food, but pig is not necessarily understood to be meat. And they have to explain how that particular animal needs to be singled out when ordering. The same thing happens in other countries with fish being meat or not meat. I think that's always something that provokes awkward interactions.

SK: One of the things we're hoping to do is think about connections between interviews. Do you know Naisargi Dave? She's an anthropologist at the University of Toronto who is writing a book called *The Social Skin: Animal Rights in India*; her previous book was on queer activism in India.

NA: I haven't met her, but I've read her work.

SK: She also brought up Hindu nationalism and the place of the cow therein. She does her research in India, so that's different, but she talked about what it's like to go back there; she's also a vegan. And one thing that stood out for me in her interview was when she talked about having meals with her cosmopolitan leftist friends in India, and how because of the articulation of vegetarianism to Islamophobia in the present moment, there's an emerging solidarity in which people are eating meat to express their opposition to Hindu nationalism. Because you brought this up earlier, I'm wondering if you are able to elaborate on any of those things. I was particularly struck by the fact you said that most or all of your family in India now eat meat. Assuming you go back there, what might that be like? Also, is there anything you want to add about the relationship of vegetarianism to Hindu nationalism in the present moment?

NA: Sure. I think we're witnessing a really crucial time for thinking about the connection of vegetarianism in India with a regressive politics that's now bordering on fascism. I was really disappointed to see that the Indian iteration of the Minding Animals conference—the largest international animal studies conference—invited Maneka Gandhi from the right-wing BJP [Bharatiya Janata Party] as a keynote speaker, given the outright violence that's being meted out against Muslims, who are often not actually slaughtering for meat but are accused of doing so, and are lynched. This is going on right now in several states in India. I can't speak so much to how to think about whether eating meat could thus be a form of solidarity with Muslims and Christians among people on the left. It is a fascinating tactic and speaks to a real attempt at solidarity. I think it also might have its own limitations. If I'm thinking of veganism as being tactical in a specific context, could meat eating be a tactical political statement in another context? Yes, but I'm not yet sure how that works in a broader time frame, given the rise in India's meat consumption. India is still the lowest per capita consumer of meat in the world, but it's growing really fast. And it's growing in part because of the economic liberalization begun by the Congress Party, and now being extended in really unequal ways by the BJP. Meat eating is growing for the elites and the upper middle class, particularly those who have access to a certain amount of transnational capital that's rolling into India. It's become a status symbol even as cow meat is singled out as anti-national. Nonetheless, I understand why symbolically some people are thinking along those lines and feeling it's important for people to create space where those dietary practices don't become a point of stigma.

But I think—especially given India's role in climate change and what it would mean to justly feed such a large and growing population—I think there needs to be a much more radical reimagining of food and relationships to land. And particularly the pro-corporate policies that are expanding India's meat export industry under the current government. Even though they promote vegetarianism at home as a kind of spiritual cleansing, India has been turned into an exporter of meat and leather. Are these industries that are conducive to an environmentally and socially sustainable future for India? That seems very questionable to me, so I don't know how it would work to try and integrate vegetarianism into a really aggressive left politics in India given the right-wing history of cow protection.

I think distinguishing cow protection from the broader question of species is one way this is kind of valid. There are distinctions: Animal rights, animal liberation, and veganism are not exactly the same thing as the domestic cow–protection movement, so how do we talk about that difference? Is that something that can be used to explore a new kind of future for maintaining what's necessarily going to be low meat consumption if we're going to address climate change? I think those are potentially important starting points.

VM: Thank you so much. A big part of why we're doing this project is to explore the role of emotions in scholarship. We're interested in better understanding the ways in which our academic interests draw focus in our daily lives and resonate not only logistically and politically but also emotionally. We've discussed among ourselves as a research group what it means for us to inhabit the role of the vegan or vegetarian "killjoy" and what an awareness of various types of violence entails for us. So my question is, given the wide scope of assemblages that you examine—for example, in *Bioinsecurities*, ranging from these large-scale political economies to the microscopic organism—do you not see violence and imperialism inscribed everywhere you look? And if so, how do you negotiate that?

NA: So, how is it possible to live and act in a world that seems suffused with violence?

VM: Yes, how do you carry that? How do you avoid drowning with it?

NA: I think that's why activism and the political work we do inside and outside the academy feels—at some points—like the only life-sustaining activity we do. For me, we can think about veganism as set of ethical practices, but it's also a set of community-building practices. We may be vegan killjoys in some moments, but then there is a whole set of practices that develop around food preparation that I never would have engaged with otherwise. I wouldn't have learned how to cook had I not turned vegan at age seventeen and needed to figure that out. So there are people who, since I spent years at some point in my life cooking with them, whenever we visit each other now that becomes part of our experience.

Our political activism is also informed by food practices. One thing we did as animal activists in college was work with Food Not Bombs, to actually feed other activists in other movements, literally during the protests, and also to feed people who were food-insecure in the community. So thinking about veganism as a political tactic rather than an ethical universal helps us think through how it can be part of anti-colonial tactics as well, the tactics that work to imagine other ecological futures even as we witness this giant extinction event that we live in.

That said, the things that you love always have to be subjected to critique, so that can feel depressing. It can be hard to be the killjoy of vegan killjoys, but asking the hard questions is important: Would it be more sustainable for us to be eating insects rather than eating cultivated soy? Would it ultimately result in less species extinction through climate change to eat the most efficient protein—which is probably insects? I have this obsession with insects. I think it's important to ask the hard questions and not to abandon critique in the face of these crises. So emotionally, yeah, it's hard sometimes. But these are also potential sites of world building, and for reinstituting futures that could have been had not colonial modernity put its grasp upon humans as laborers and animals as nature.

CHAPTER 11

Interspecies Intersectionalities

Harlan Weaver

Harlan Weaver is a professor of gender, women, and sexuality studies. His research stems from the intersections of transgender theory, affect theory, critical race theory, and human–animal studies, to name just a few starting points. A doting pitbull owner and advocate, Weaver discusses how to understand and communicate identities and experiences in ways that are authentic and ethical. Working to unravel how identity formations not only co-exist but mutually co-constitute, Weaver shares his experiences of feeding himself and his canine companions while navigating convoluted foodscapes. We interviewed Weaver by Skype on September 23, 2016.

SAMANTHA KING: Thank you so much for joining us; we're really excited you agreed to participate in our project. We'll begin with the biographical: when and where you were born and raised; your formative, cultural, intellectual, and political experiences; and how you became an academic.

HARLAN WEAVER: I was born and raised in the San Francisco Bay Area. And I grew up at a time when the school districts were some of the worst in the country. So lots of my peers, white peers in particular, went to private schools, and I grew up in that intellectual culture, which gave lip service to a lot of social issues, with questions of feminism at the forefront—but in a way that was still very white. But I basically started off as deeply committed to feminism because I had a lot of rage. Then I went to college at Wesleyan in Connecticut, which was amazing; I loved it there. I majored in what was then "Women's Studies" and is now called "Feminist, Gender, and Sexuality Studies." And part of growing into that community for me involved collective cooking; that's really when I learned how to cook and I mostly only learned to cook vegetarian food, because that's what we ate. That's where it started for me; so much for me of the good thinking and community building I've done has happened and continues to happen in the space of a kitchen. Which came in handy again later because my advisor in grad school at History of Consciousness at UCSC [University of California, Santa Cruz] was Donna Haraway. Donna, and pretty much all of Donna's advisees, loved to cook, so we did a lot of cooking together and hanging out in the kitchen, and I still cook a lot. Also in addition to grad school, in my twenties and thirties, my cultural milieu was the queer community in the San Francisco Bay Area. The specific lesbian and trans cultures that I lived in were really activist-oriented. So especially during that time and in that community I witnessed a lot of key conversations and transformations, and understandings of community around what have been called the "border wars," which was basically, How do trans folks fit into these communities? Are people leaving the communities when they transition, etc.?—debates which continue, in many ways, to this day. But then, what I saw was the active creation of space for trans-masculine folks in what had formally been mostly lesbian and self-named Dyke Bay Area spaces in a way that was a really positive political development. However, those spaces were not and still are less welcoming to trans-feminine women; there's a lot more work that needs to be done there. I also spent a lot of time volunteering in animal shelters during that time, and part of that was I adopted my own dog, Haley. And shelter work, it's extremely hands-on, intense physical and emotional labor. So even with all its troubles and flaws, I developed a deep connection to animal rescue politics in particular.

SK: Did you go to work with Donna Haraway because you knew you wanted to work on questions related to dogs or animals?

HW: Oh no! I went to work with Donna Haraway because I got into Hist Con! And because my favorite essay of all time is *Situated Knowledges*.

I went to work with Donna because I was obsessed with thinking about knowledge politics and the relationship among knowledge politics, experiences of embodiment, and questions of identity. Of course, these issues come up a lot in animal studies stuff; for example, one of the key themes I see folks wrestle with is the question of how to put forward ethical knowledge about others and their experiences without further othering or erasing those others. Some theorists try to answer this and related questions or representation through the idea of "speaking for" animals, but, for me, "speaking for" involves a form of erasure that I dislike. In addition, "speaking for" prioritizes metaphors of speech in a way that ignores how nonhuman animals themselves communicate. But I was lucky in that one of the neat things about working with Donna was that I got to hang out with folks who came to work with her, and so I was introduced to Maria Puig de la Bellacasa, who shared her sense of "thinking with" with me when she came to Santa Cruz. As I got to enjoy hanging out with and, perhaps unsurprisingly, cooking with her, I came to think about a different approach that rested on a different preposition, "with," rather than speaking or thinking "for" or "about," which introduced a different kind of proximity and a sense of solidarity that I felt avoided the problems of "speaking for." So, for me, coming into contact with folks working on those and similar questions was the main draw of working with Donna: thinking about philosophies of knowledge and eventually onto-epistemologies, which then came up for me more later when Karen Barad came to Santa Cruz.

SK: What do you see as the primary purpose of your academic work, and why do you do what you do?

HW: I have two dual themes going on; the first goes back to questions of trans embodiment. In grad school, I became really interested in combating the ways that feminist and queer theories, in particular, misunderstood and misread trans folks. One of the main critiques that a number of folks have made, and which also surface in challenges to Judith Butler's *Gender Trouble*, is that trans bodies are made to stand in and do theoretical work as figures of resistance to gender norms, for example, in ways that disregard or elide the lived vulnerabilities that make trans lives so transgressive in the first place. My dissertation was about putting forward a different knowledge politics, or, borrowing from Barad, an onto-epistemology, rooted in an understanding of trans experiences in a way that was centered on how trans folks understand themselves, rather than understandings focused through the lenses of how nontrans folks perceive them—which were often really about questions of passing, and cultural and social legibility.

And then as an owner of the pitbull-type dog with clipped ears, Haley, I became very aware of the stigma around pit bulls. People said horrible things to us all the time. And granted this was in the early 2000s, and much has changed in terms of my day-to-day experience as an owner of the pitbull-type dog; I don't tend to experience the kind of overt hostility now that I did then with my current dog, Annie, who, by the way, is much more of a handful than Haley ever was! In fact, now people tend to bring up the differently problematic trope along the lines of, "Oh! It's about how you raise them," reinforcing a troublesome nature/nurture divide. But before, in my early days with Haley, people would just literally not talk to us, or spit at us, and/or curse at us. So when I had Haley, I became very alert to the ways that stigma shaped her life and the possibilities that dogs like her even had, perhaps especially in terms of their prospects in most animal shelters; so many dogs are euthanized just because they're pitbulls or pitbull types. And then there is the problem of BSL, or breed-specific legislation; we would drive places and I would have to literally look on a map to make sure it would be okay to let her out to pee.

So my experiences with Haley got me thinking about dogs and the experiences of dogs labeled as dangerous, and one thing that really stood out to me in these worlds were the kinds of connections that were made with race by both activists and anti-pitbull people. So pro-pitbull folks would leverage this language of "pitbull underground railroad," and "canine racism," and then anti-pitbull folks would also use messed-up racialized language about how these dogs are always owned by "thugs," and "gangsters," along with other racially coded terms. Or they might introduce problems of pitbulls and "white trash." Those ways of framing and connecting pitbulls to race, and specifically black and brown masculinities, really pushed me to think about how to consider first race, and then, later, questions about sexuality, class, and gender, in the context of these discourses. So as part of my thinking on that front, I began by countering what I think is deeply problematic in a lot of earlier animal studies work: where race-related analogies loom large as ways to do animal advocacy. So, for example, you have books like Marjorie Spiegel's *The Dreaded Comparison*, which pairs images of enslaved people and slaughterhouses. To me—and I think a lot of folks feel the same—I see such materials as appropriating images and language from human slavery to make arguments on behalf of animals. And when I talk about animals and race, I want to *not* do that kind of politically and intellectually problematic and appropriative work. Instead I try to do careful work based in the concerns of critical race studies and queer and gender and sexuality studies in a way that doesn't borrow or appropriate language from marginalized peoples but rather thinks through these troubled connections in ways that foreground concerns of, say, critical race studies, *together* with those of animal studies.

Part of my work in trying to think through the concerns of marginalized humans and animals together is to counter the analogical or parallel thinking that happens about race and gender in this type of animal studies work. For example, speciesism was initially, and continues to be, defined through analogies with sexism and racism, as being akin to them. This sets them up in parallel, in a kind of simile, by using these other ideas to build up the idea of speciesism; because of how that "formula" of analogical thinking works, you end up putting them in parallel rather than in connection. In a broader, epistemological sense, I feel like those kinds of intellectual moves make it impossible to think questions of species *together* with race, gender, and sexuality because they're held apart by the very nature of the formulation. So part of what I do is about producing a different analytical or theoretical framework where the concerns of marginalized humans and nonhuman animals are conceived of *together* and *through* each other, rather than apart.

VICTORIA MILLIOUS: That's great, thank you! You've addressed this a little bit, but how did you come to write about pitbull politics?

HW: It was really through adopting Haley and experiencing the stigma that was leveraged against her. This was not just what would happen during our daily walks, when people would cross the street, say messed-up things to us, throw things at us, etc., but also through our larger movements through space, as, for example, when we would drive to Montana to visit one of my friends over the summer and I'd have to be really careful not to stop in locations with BSL because I didn't want her to be seized and euthanized. And I noticed the profuse connections with race I outline above, but I also began to think critically about animal rescue. I was struck, for example, by the fact that people don't ever put air quotes around *rescue* when they're talking about animals. One of my favorite interlocutors from my fieldwork used to say, "Unless you're running into a burning building, you're not actually rescuing them." I agree, in part because with rescue you're most often taking animals out of the supposedly benevolent hands of the state that was first trying to aid them and then trying to kill them through a pretty curious interface between animal shelters and grassroots/NGO-type "rescue" groups. So I became curious, witnessing this, about how animal rescue, however unthinkingly, inherits a complex history around salvation and saviorism that has so much to do with whiteness and histories of settler colonialism. More specifically, in the rescue worlds I saw folks inheriting these histories that they were completely unaware of through their repeating the rhetoric of salvation. And then I would also witness people in animal shelters routinely make comments such as the animals were "in jail," or "on death row," drawing from prison-related language with no attention to vast injustices of the prison-industrial complex. This language and understandings, in conjunction with the other race-related language I note above, both troubled and fascinated me, which is how I came to write about pitbull politics.

VM: Can you describe a key concept or idea that guides your thinking about multispecies relationships?

HW: My main concept is what I term "interspecies intersectionalities." In it, I draw from Kimberlé Crenshaw's early writings on intersectionality and the ways that that has been taken up by many people, and pair that with human–nonhuman animal relationships in particular. My basic thinking here is to try and understand how relationships between humans and animals don't just reflect but actively *shape* experiences of race, gender, sexuality, nation, species, and breed. I've been taking that concept and, through my book project, pairing it with a term that I'm borrowing from Donna Haraway, which is *multispecies justice*. But I am doing this work with my own twist, in the sense that I'm interested in thinking toward, or thinking with, a kind of multitude which is very much about producing a theoretical framework of complex overlap and intersections and mutually implicated ways of knowing and being that counter the parallelism and analogical thinking evident in the framework of ideas such as speciesism that I mentioned earlier. My work is really about: How do you think marginalized humans and animals together? Lori Gruen is helpful for me to think with here, for she introduces, in a brief piece for *The Huffington Post* on Cecil the Lion, that the central problem is about zero-sum thinking. She notes that in the uproar following Cecil's killing, people tended to ask questions like, "Why do people care more about animals than they do about impoverished African humans?" For me, one of the main interventions that emerge from my pairing of interspecies intersectionalities and multispecies justice is giving a framework that does not do this kind of zero-sum thinking where there's not enough care, where you're pitting one commitment against the other, instead of asking where these commitments—to animal welfare and racial justice, for example—are coming from, in a way that enables us to better understand and think them together.

VM: I look forward to reading that book.
HW: Me too, right?

SK: Well, in the meantime we've been enjoying reading your various articles. In fact, the next question is about *The Tracks of my Tears*: Could you talk about the resonances you've seen between trans species and trans affective politics?

HW: I can try! I just participated in an interchange for the "Tranimacies" special issue of *Angelaki*, which Eliza Steinbock, Mariana Szczygielska, and Anthony Wagner are editing. They asked me to do that, and it was really hard! So, for me, trans affect comes from producing better understandings that emerge from thinking with affect and thinking of the ways that affect trans folks' experiences rather than any questions of cultural legibility, etc. One of the key passages that really speak to my thinking on trans affect is a conversation between Jennifer Finney Boylan, the professor and famous novelist, and one of her best friends, Richard Russo—another famous novelist—which Boylan relates in *She's Not There*. She describes her move to transition, and he replies something along the lines of, "Well, thanks for telling me, and when did you make this decision?" And she responds, "It was more of an erosion than a decision." I think of that as an example of trans affect, as a ceding to a bodily movement, rather than, "Well, I'm going to decide to put on some different clothes tomorrow." It's also notable because one of the things you'll notice if you start reading a lot of trans memoirs, as I did, is up until probably the late 1990s they all sound the same. This is because there was, and still is, a lot of medical gatekeeping, where you have to present with this particular narrative of being "trapped in the wrong body" for a long time. You also had to identify as heterosexual. You had to use that specific language adamantly, and if you didn't, then you didn't get hormones or any other kind of treatment. So in order to medically transition, people ended up telling these stories and I think—probably as expected—internalizing them, even if they weren't necessarily true to their experiences. So one of the ways I tried to intervene in my dissertation by reading for trans affect was to examine how there is still stuff going on in the confines of language that seems, on its surface, very limiting. This is to say that I saw differences emerge in terms of ways of knowing and understanding embodiment in spite of the sameness of the language required by medical gatekeeping of being "trapped in the wrong body." This is the sense of trans affect I'm working with. And this is where I struggle to connect it with trans species thinking—there is a certain kind of prediscursiveness to it that invites a lot of the critiques you hear of Deleuzian-style thinking, that in thinking with affect and bodily movements you are, in a way, unmarking bodies and therefore failing to address intersectionality. This is the struggle!

Because of this issue, I may not have very good resonances for you, but in terms of the trans species politics, one of the ways to bring in trans affect would be to take up interspecies affects in the ways that they both shape and reshape experiences of gender, race, sexuality, species, and breed. So, for example, we can talk about the specific kind of love that gets invoked in dog rescue cultures as something that comes out of a whiteness that is itself shaped by inheritances of colonization and colonial thinking. And that kind of human–dog love and dog rescue all inflect and then reshape that whiteness in ways that can be hopeful or not. And then you can bring this understanding to bear on the more troubling instances when, for example, you have folks who come to pay attention to species-specific injustices that they may not have noticed before, but they'll do it through the appropriative mechanism I noted earlier, such as one of most striking images I found recently on Facebook of an "All Dogs Matter" rescue group.

SK: Oh God.

HW: Yeah! So there's this emergent trend where, in plain speak, they kind of "double down" on the whiteness. They go into this interspecies relationship and emerge even more firmly entrenched in their resistance to recognizing racial injustices. So that's one way to approach the question of resonance. But then there are also the transient kinships of animal shelter life where there are these moments of hope and ephemeral, yet promising, connections that happen in the broader expanse of very normative and troubling narratives. So in an animal shelter you have folks having really interesting moments and interactions with dogs, but they then leave. There is this kinship and affect happening in really interesting ways, but in a larger context of grooming these dogs to hopefully be adopted into what is an implicitly white middle-class "forever home," with picket fences and all that. And this is because people have very set ideas about what makes a good home, that are very deeply shaped by class, politics, and the resources involved: whether you have a back yard or not, whether you own your home or not. Most shelters I know of—and this is also the shifting landscape of pitbull politics—won't adopt out bully breed dogs to people who don't either have a lease explicitly stating that they can have a bully breed dog, or own their own homes. And in a lot of urban areas that's pretty much impossible. But within this broader landscape and aim of shelter life, you still have these really interesting moments of touch and intimacy.

Then the thing I've been thinking with most recently—have you read the new Vinciane Despret book? Or her *Embodied Empathy* essay, which I love? She writes about ethologists who use their bodies to understand animals. She talks about how Farley Mowat, for example, ate mice so that he could answer the question of whether and how wolves could survive when caribou migrate and they don't. So he ate mice, which was his way of asking, through his body, this question. Although I should note that, according to Wikipedia, people don't necessarily believe the Mowat version of that story. But Despret has helped me think with and understand the importance of bodies, human and animal, in interspecies relationships; the idea of using your body to better understand nonhuman animals that Despret highlights is super interesting to me.

Building from Despret, one of my favorite dog trainer writers is Patricia McConnell; she's also an ethologist. In her dog training work she does these seminars, and one of the examples I really like that she gives is trying to teach a dog a good sit-stay. She'll bring humans up on the stage and press into their space, without actually touching them—she just moves really close to them. And if you try it among yourselves, you don't want to move! She uses that to make larger points about how, when you're practicing dog training stuff, you should try it on a human first to see if you get the body timing right and then work with your dog on it. It's really neat 'cause I've done it and have been like, "Oh! This feels really weird; now I know how Annie would respond to this technique better." So in this, I see a bodily epistemology that's also deeply affective because it's really about bodily movement and how you feel as an embodied being, which is itself part of a larger ethics of positive reinforcement–style and, I would argue, a pretty feminist, dog training approach. And yet, those kinds of teachings are not very widespread. Folks have heard of Cesar Millan, but not many have heard of Patricia McConnell. However, McConnell's stuff is really fascinating to me because there's this bodily thinking going on in her approach that traverses species lines in a way that's deeply affective. And yet, and this is a big yet, her type of approach is also undergirded by the fact that you need certain resources to be able to access that knowledge at all, and you need to be in a space where the norms that she's teaching dogs toward are norms that are possible, and regarded as norms. If you're in a space where, say, it's impossible to take your dog out on a leash walk, it doesn't really matter whether or not they behave well. And, getting more critical, we need to examine what "behaving well" means for different contexts.

Overall, my sense is that tracking affects in the ways that resonate with the knowledge politics of working from trans lives comes out of attending to those affective emergences that result from, for good and bad, interspecies relationships in the worlds of rescue in particular. That's where the resonance happens. Then we can follow this by devising a politics that builds on the affects and the nonproblematic aspects and/or the more hopeful worldbuilding moments, those that don't necessarily feed into or inadvertently inherit these histories of oppression and domination, and then promulgate those more promising affects and politics through practices and intellectual interventions; this would be where and how to move that resonance forward. That said, I admit that this connection between trans affect and interspecies intersectionalities is still a pretty tenuous one in some ways.

SK: It's provocative how you're challenging the kind of affect theory that is very detached from intentional thinking, and also challenging kinship theory and normative modes of kinship, while at the same time moving trans theory in new directions. It's impressive that you're holding all of that together.

HW: I think it's seeping. Seeping would be a very hopeful description. But thank you, I appreciate that. I do think that bodily knowledge stuff is really neat especially when you think about the politics of dog training, because *there* is a world divided. You have the old-school trainers versus the positive-reinforcement types and the positive-reinforcement methods; in my reading, a lot of bodily thinking goes on in the latter. Which is interesting because those methods come out of behaviorism and classical and operant conditioning, which are notorious for the "Black Box" approach, where all you're interested in is some cue or reinforcer that will make that next behavior more or less probable. So on the surface it wouldn't seem that positive-reinforcement trainers would be at all interested in how animals think, and yet all of those folks are deeply interested in how dogs think, and the ways they go about conveying that to their human audience is just fascinating.

SK: It's been quite helpful for me to read. We're going to change tack a little bit here: Could describe your relationship to eating animal products? Has it been a preoccupation for you? How have your feelings and thoughts changed? And then I was also thinking about your notion of "becoming in kind," which is really provocative. Whatever you want to tell us about that.

HW: I think about food a lot—mostly because I love cooking and I find it really soothing. My students and I have been talking a lot about how the world is falling apart right now, and we had this whole moment yesterday where we just wrote on the board all of our coping strategies. And one for me is cooking, and listening to podcasts. Together. Now I get Internet in my kitchen so I can also watch Netflix, which I didn't used to, so it's shifted a little bit, but the podcast is more soothing.

SK: Well, you must have really good knife skills, if you can watch Netflix and cook at the same time.

HW: I rewind a lot! Anyway, I do think a lot about food. I think less about vegetarianism and veganism than I used to, which is interesting because I find that, while they are really common in the rescue world, they are much less so in the training worlds I frequent. But I am an omnivore; I was vegetarian for a long time and then I got Haley and I was feeding her these pig ears and stuff, and I was like, "Oh. I can handle this," because, through her, I came to find meat a lot less gross. And then I was dating someone who was really not into vegetarian stuff, and so I gradually shifted over.

In terms of consumption of animals, I do think it's always complicated. I recently taught this really neat piece by Erica Weiss, "There Are No Chickens in Suicide Vests," which talks about the shifting politics of veganism in Israel. And then we read Claire Jean Kim on Holocaust and slavery analogies and vegan activism, and as part of that I showed my students a video from Dr. Breeze Harper, who is a friend and colleague; she does the Sistah Vegan Project and I like her work a lot. And this sequence helped all of us think through how one of the ways food is so complicated lies in the reality of economic constraints, and the ways those have to do with the politics of whiteness and class, in terms of what you can afford to eat. There's also geography; right now I live in the middle of Kansas and it's hard to find good produce period, much less ethically produced. Then in a more body-wise way I have two really severe food allergies, to sesame seeds and tree nuts. And during grad school I developed a lot of food intolerances, so I spent seven years not being able to eat wheat or dairy and then I did this alternative treatment based in principles of traditional Chinese medicine that actually worked—which I learned about from Mel Chen. That treatment got me thinking about traditional Chinese medicine—which is everywhere in the Bay Area in a lot of ways—and so I began to build an understanding of food as medicine from that. So if you have a cold, eat a lot of daikon and onions to clear your lungs, for example. So for me, this added layer of epistemology, of understanding the nature of food, productively complicates the politics of eating. And, in sum, I think there are a lot of different ways to read food that come out of that mix.

But again, for me, basically it's complicated and I generally do eat meat, and just try to be super honest about it: I eat nonhuman animals. But I also try and do so with an eye to the human and nonhuman labor in all its forms involved in that, as well as my own economic and time limitations. Right now where I'm living, there are a bunch of really neat local farmers who will sell you half a pig. I don't have a refrigerator that large, but I'll buy stuff from them. One of the things I think about in refusing to get pigs from factory agribusinesses is that it's not just because of what they do to pigs, but also because of the recent history of ICE [Immigration and Customs Enforcement] raids on slaughterhouses, which are very troubling. And then there's the union busting that happens in those spaces as well. And so I try and concern myself with the ways humans are marginalized and oppressed through an entwined history and present of racialized capitalism and classism in those spaces, which you can see very clearly in Timothy Pachirat's *Every Twelve Seconds*. There are also other problematic norms produced and reinforced there. My class actually watched one of those pig farm hidden camera videos the other day, and one of the things we talked about was how there is this hidden culture of toxic masculinity that the video revealed, which we then tied to not just the local factors of that particular labor and space but also the larger framework of racial and class politics that demand and produce that specific kind of masculinity through that setting.

So, to go back to the pig farmers I know here, they are doing something differently, and yet that difference is made possible by certain resources and investments. Many of the "back-to-the-farm" folks are generally not folks that actually grew up in those areas but are white folks who transfer their capital there from other venues. I mean, it's expensive to get land, and to buy a farm, and you end up investing a lot up front. So there's the ways that folks who are entering these farming movements anew are doing so using inherited capital that comes out of centuries of oppression involving slavery and settler colonialism. There's also, for me, a deeply troubling whiteness, and differently troubling gender politics, that goes into and makes possible spaces like farmers' markets and CSAs [community supported agriculture], that I would say is actually of a piece with the problematic racial, class, and gender politics of agro-business. The dynamics of both are troubling in different ways and in ways that actually have similar roots. You can see it with the strawberry boycott right now. Indeed, one of the other interspecies intersectionalities, and becomings in kind, that I think is really neat to think with in this regard, is thinking with plants. So I am interested in examining the kinds of identities that emerge from strawberry picking in particular conditions, along with specific kinds of racial and class experiences. All of which is to say that, for me, the politics is more complex, and in a very productive and fascinating way, than "vegan or not vegan."

VM: In relation to consuming nonhuman animals—do you want to tell us about what you used to feed Haley, what you feed Annie, and how you make those decisions?

HW: That was my question for you! Because I spend so much time thinking about what I feed my animals, and I thought, "How did this not come up?" That question I love; it's something people don't talk about that much in these venues.

VM: I am also the resident dog sitter; I've dog sat for a variety of faculty in the department and it's just something always on the brain. The food, the feeding routines, the treat protocol conversations. . . .

HW: Yeah! Haley, like a lot of short-haired dogs, developed really severe food allergies when she was around seven, and the only thing that worked was to feed her the raw food diet. We worked with a really wonderful holistic vet, Dr. Cheryl Schwartz. She was one of the first in the country; she wrote *Four Paws, Five Directions*. She helped me develop a diet that was attuned to Haley's bodily needs, in terms of reading her body through the lens of traditional Chinese medicine. It was interesting because I would only go to Dr. Cheryl very occasionally, for super long-term health stuff, because she was very expensive. And then we would go to our other vet who would excoriate me for feeding Haley raw meat, and

I constantly felt caught between these two contradictory epistemologies in terms of understanding what was good for her. One of them that was very much shaped by industrial agribusiness, and the other shaped by what I knew, which was that she looked like she had hives if she didn't eat this particular food. The other vet was very expensive also. And then sourcing that food was its own experience, trying to find a good source of goat meat led me to spaces I hadn't been in before. So we ended up doing a lot of Halal meat shopping because that was where I could get the best meat for her, and where I was careful not to tell people that I was buying it for my dog, for obvious reasons. In some ways I feel like, when I have the money, which I don't always, I put feeding my animals well before I put my own feeding—like many of us, no doubt! The other thing I've been doing with Annie, who, fortunately, has champion digestion; it's amazing—if we were all so lucky!—is that she eats her meal every day through training. I put all her kibble in a treats pouch and we practice tricks, or go around and try to be okay when other dogs walk by—which isn't always very successful. So our food relationship is really about a much deeper positive training relationship. It's one of our ways of speaking and communicating; like, look, good things happen to dogs when they sit down. It's basic operant conditioning, but leveraged for the greater good of her and our relationship.

VM: I commend you on that; there's a lot of effort there. I'm glad you mentioned the extent of the cost and what's involved in thinking through all these things, and then the practicalities—especially while working in the precarious labor environment of the academy.

HW: Yes! Sometimes with Annie it's just learning how not to bark at dogs on TV because we're really tired and we don't have energy to leave the house.

VM: We've discussed this a bit, but we're wondering if you could discuss if and how your writing about pitbulls, racialization, kinship, transgender affects, or other subjects has shaped your dietary practices or vice versa.

HW: Definitely not vice versa, yet. I can see it probably coming up when I get to the really amazing chapter of my book where I describe multispecies justice and interspecies intersectionalities together in all of their glory, but I haven't written that yet. But in terms of the writing about pitbulls, not any more than we talked about: It makes me think more and more about how it is so complicated and how there is no black and white. You know, there are so many different ways to be unethical when eating. I've been bringing my students chocolate because our class happens right around the time everyone's sugar low hits, after lunch. And I've been thinking, I need to make sure I'm bringing chocolate that doesn't involve the problematic labor practices, the slavery-like conditions that I've read about. And it's generally vegetarian—it's not vegan—but even so. It just keeps getting complicated. And sometimes it feels like too much and other times I feel like, "Okay I can take it on!"

VM: Do you have a story about a time when your dietary practices have been the subject of awkwardness, celebration, or hostility?

HW: When I couldn't eat dairy, or wheat, or sesame seeds, or tree nuts, I had a campus job talk once, and they were like, "Let's go to this really great dinner place!" And if we had gone there, literally the only thing on the menu I could have had would have been the fifty-dollar steak. Which you don't order when you're a campus visitor, right? So I asked, "Would it be possible to go to that Thai restaurant just down the street?" They were super nice about it, but it was one of those things where you don't want to be the difficult person when you're applying for an academic job. But also, cooking is always a celebration in some ways; I've had so many dinner parties. When I was at Wesleyan I did the ASI fellowship there, which was wonderful, and my friend and colleague Juno Parreñas and I rented a house together and we would always have everybody over and cook together. We always just defaulted to vegan; we made a lot of sushi that summer. It was also really hot, so it suited in many ways. This goes back to Donna's thinking on species and messmates—on eating together as a way of building kin, intellectual kin in that sense. But I also have a lot of really good friendships from that period of time.

SK: Is there a key dilemma or question that haunts you?

HW: I think it's the one we talked about: How do you really, in a strong way, take on affect theories in the ways they've been taken up by so many folks, and thinking with animals the way that I think Vinciane Despret does a marvelous job of, and at the same time build in a critique based in interspecies intersectionalities? So my critique of Despret and Patricia McConnell, eventually, is that dogs can be racist and that much of the practice of rescue and training inherits and reinscribes legacies of racism and colonialism, and so I constantly wonder how to talk about this sense of deeply affective bodily knowledge building, and at the same time talk about race and class and gender formations that come out of that, because of this prediscursive or unmarked way that the affect theory can travel. It's not remotely easy to bring it together with those very legitimate concerns about whiteness and rescue politics. So that would be my dilemma. Maybe someday I'll fix it, and maybe it's just not doable, which is fine too. I feel like if there's anybody who has a clue how to do it, it's probably Sara Ahmed, so I need to reread more of her work.

VM: In *Becoming in Kind*, you write, "I'm not positing that pitbulls are themselves racialized, a mood that ignores disparities and histories of violence and species." Have you come across work that draws problematic moral equivalencies between racism and breedism?

HW: I would say I come across it more in the *world*. I mean, I hear it all the time. I feel like I've done a really good job recently of reading people who don't do that: Claire Jean Kim's stuff doesn't do that at all. My class read some of Susan McHugh's work that talks about the sled dog massacre in a really remarkable way. But the problematic equivalencies do come up constantly in the activism I witness. In fact, one of the reasons I do the work that I do is because there are so many ways that we relate with animals that *don't* have to do with eating. So I don't read a lot of stuff about the food, about veganism in particular, and I think that's where a lot of those comparisons happen in intellectual work. I think if there were more people writing about training and pitbulls and stuff you might see more of that pop up, but I think I'm friends with all five of us. There just aren't that many people doing it. I think if it were a more pervasive topic, I probably would have seen it.

SK: That's one of the reasons why we're trying to talk to people who don't use veganism as a starting point or endpoint for analysis. It is interesting that many of our interviewees have turned out to be vegan, but we didn't know that going in and it certainly wasn't a criterion for participation.

HW: You know what's funny, is I met Lauren Berlant at the SLSA [Society for Literature, Science and the Arts] in Europe, and she's such a neat person, in addition to her thinking, which is fabulous. I asked her, "Why aren't you an anthropologist?" Because she was so curious about everything and she was like, "Well, you can't be an anthropologist if you're vegan. You can't turn down food as an anthropologist."

CHAPTER 12

Living Philosophically

Matthew Calarco

Matthew Calarco is one of only a few continental philosophers who write explicitly about animals as food. Calarco describes his long-term veganism as a set of daily practices that call into question natural attitudes about meat eating. A key concept for Calarco is "indistinction," which he describes as a purposeful blurring of the boundaries between animals and humans that enables alternative modes of living with, relating to, and being with others. Here Calarco discusses the shared meatiness of humans and animals, suggesting that we are both meat and more than meat. Calarco embraces an openness in his scholarly work and concludes by sharing his hope that the field of critical animal studies will similarly resist adopting a closed framework, instead remaining open to new discourses and traditions available through intersectional scholarship. We interviewed Calarco by Skype on July 27, 2015.

SAMANTHA KING: Can you begin by telling us about where and how you were raised, and your formative political and intellectual experiences?

MATTHEW CALARCO: I was born in 1972, in Escondido, California, which is not far from where I'm currently living and teaching in Fullerton.

Probably the only thing remarkable about where I grew up, in Escondido, is how conservative it is. So I was raised in a conservative milieu, but bizarrely, at some point in my mid-teens I started getting interested in questions concerning a whole bunch of social issues: white supremacy, racism, colonialism, sexual difference, and gender, but also my interest in animal and environmental issues started to sprout at this time. I've been thinking about this and trying to figure out where these interests came from, where the passions came from, and I have to be honest—it remains pretty opaque to me. As I was thinking about this question, I was reminded of Nietzsche's quip in *The Genealogy of Morality* where he talks about his anti-moral, anti-theological disposition. He says it's a kind of *a priori*—it's always been with him. And I would say those passions have always been with me in a strange way. The only thing I can think of that pushed me in that direction is my family. They were not conservative, but they weren't exactly radical either; they always encouraged me to think critically and freely, and so I think I felt like I had the space to explore those issues.

I was reading philosophy at a very early age and just had no idea what it was. I didn't know the field; I didn't even know it had a name. But I'd always been deeply interested in what you might call big, existential questions about the meaning of life, and how we fit into the bigger scheme of things. So I went into college thinking I was going to do law, or something like that, because it seemed it would fit well with my interests. And I took a philosophy class by chance, and it was like "Oh!" I had no idea there was an actual field dedicated to the kind of questions I was interested in. I was lucky—I don't know if you know Andrew Feenberg? He's a big thinker in critical theory of technology. I had him as a professor in my freshman year, and I thought, "This is it, this is where I belong!" But I was so naïve, I had no idea how philosophy was actually structured and that some of those issues we had touched on weren't central to the discipline at all. Philosophy in the United States in particular, is *completely* dominated by metaphysics and epistemology, by questions about the nature of reality and the theory of knowledge. And so the concerns that I had were *not* central, and I found that out to my surprise as I moved through the field. I decided to major in it anyway, and to stick with it. I was told repeatedly, "Hey, the stuff that you're interested in, you really don't belong here. You should be in another field. You should be in cultural studies or women's studies. Philosophy isn't really about these kinds of questions." But I had a kind of passion, and this stubborn insistence on staying inside the field and insisting that those questions fit.

SCOTT CAREY: What is it like to study animals in a philosophy department?

MC: Very strange! Let me start in reverse order. When I was hired at Fullerton, where I currently work, I was welcomed with open arms. They wanted me to focus on what I was doing; they thought it was important and they made room for it. Moving backward, I found grad school to be a strange experience. I went to SUNY Binghamton because Maria Lugones was there, and a whole bunch of other decolonial thinkers, and I was interested in that material. Even though I wasn't going to write on it, I wanted to study it and work with those folks. But there was also a pretty solid contingent of continental philosophers whom I wanted to work with, who studied Derrida, Heidegger, and Deleuze. I wanted to introduce animal issues into that context, and, to be honest, I received considerable pushback. They thought it was a bizarre topic. None of the people on my committee or in my department were convinced that animal issues were really of much interest in continental philosophy, or in philosophy as a whole. So I was discouraged from doing it. Things have changed culturally, and the conversations have changed, but my entire time as a student, it was a constant struggle to insist that these were important issues, the animal ones in particular. So philosophy has been—as it is with most of these things—recalcitrant and really stubborn, and you have to continue to prove the relevance of whatever you're interested in if it falls outside of the realm of mainstream questions and divisions of the field.

SC: Much of what you write about in your work has to do with the idea that humans and animals embody consumable flesh, yet we live in a context where we learn to deny that meatiness of ourselves. Do you see helping people come to a better understanding of this notion as a philosophical endeavor? Or are there other ways in which this process of learning takes place?

MC: It happens along multiple registers, and in multiple domains. As a philosopher, I see myself as only working in one small segment of this bigger project. When I am looking specifically at the embodiment issues, or meat, or flesh, it's odd—I find myself almost *always* pulling from nonphilosophical sources. From literature, from art, poetry—whatever it might be. There's very little inside the philosophical tradition that wants to think about those things, so it's a matter of philosophy's being dependent upon something outside of it to generate a lot of these questions. What philosophy can do, and what philosophy has to say about these issues, is somewhat limited.

SK: Could you describe how the field of critical animal studies has changed over the time that you've been engaged in your work?

MC: It's changed a lot! It wasn't even in existence when I was first working on these issues. When I was a graduate student in the late 1990s, there *was* no field of critical animal studies. There was animal ethics, and also the field of animal minds, inside philosophy. In both of those instances, it's the same ball game: If it's animal ethics, it's "Whatever's important in ethics—do animals have it?" and in animal minds, it's "Whatever it is that constitutes human minds—do animals have it?" That, to me, wasn't of much interest. I had a whole lot of things I wanted to investigate, and I was struggling to find a way to get that work out, since it did not fit squarely inside animal ethics or inside animal minds. I was also reading outside of philosophy at the time, in history, anthropology, literature, feminism, and biology, and saying, "Oh man, this is more of what I want to do." But the problem with that work is that it was isolated; it tended to be the province of only a handful of individuals, and it was removed from the kind of ethical and political concerns that were salient to me. So I was trying to help create the space for some kind of critical animal studies. And that started to gain prominence, with the help of lots of folks, in the mid-2000s. And at the time when the term *critical animal studies* came about, there were lots of other terms floating around too. Some people wanted to talk about animal studies, or society and animal studies, or human–animal studies, and those terms still float around and still fluctuate. I guess my work would be most associated with critical animal studies, and I'm fine with that alignment. I think I am closest to critical animal studies, which is typically understood to be the more directly political and engaged aspect of animal studies. And most of my dearest friends and most precious intellectual and political allies are in that field. So that's where I think I most belong.

SK: What do you see as the promises and pitfalls of the field as it stands today?

MC: Well . . . there's both. The emergence of critical animal studies as a field has been a wonderful thing in all kinds of ways. The stress on ethical and political issues when one is looking at animal questions; the idea of putting strategy questions at the foreground, and not limiting ourselves to mainstream liberal, legal, and policy strategies; looking at different direct action tactics; the creation of organizations, forums, venues—all those things are done by scholars and activists in the field. The Institute for Critical Animal Studies—ICAS—has done great things along those lines. When I was a graduate student and when I first started out in the field, I would have never dreamed that this kind of organization and discourse and set of practices would have emerged so quickly. To me it's a wonderful thing. At the same time, as with any organization and anything that tries to be a discipline within the academy, it's going to be domesticated by tendencies within the academy. And there are also the kinds of problems that happen with any organization in terms of internal struggles. There's a lot of debate about which sources or intellectual traditions are proper to the field. There are people who say, "No, this is an anarchist thing" or "No, this is a critical theoretical thing"; "Should it be pulling only from analytic ethics?"; "Should it have veganism as a moral baseline?"; and "Who belongs to our group and who doesn't belong to our group?" Those kinds of debates are problematic, and certainly not my cup of tea. The people who are having them, though, are goodhearted, and care about justice and improving the lives of animals, and linking animal issues to a whole set of other struggles. And over time, as those things get hashed out, it'll work itself out in a good way. But those are some of the pitfalls, as the attempt to shore up an identity—and this is what happens with several political movements—becomes ultra-exclusive at the same time.

SK: You've described your approach to eating animals in your written work. To us, your work is exemplary in this regard, because it shows the utility of theory for wrestling with practical and political issues, and vice versa. I've been struck by what seems like something of a disconnect between theory and practice in some work in the field. Could you review your approach to consuming animals for us here?

MC: Yes, and by the way, I share your feeling of a disconnect. So I don't know if you've hung around a lot of vegans, but this happens a lot as you talk to each other and tell these, they're almost like conversion stories. In my case, in my teens—the general period I talked about at the beginning—I started reading about the factory farming of cows, and chickens, and pigs. And, I'll be honest, I was shaken, I was deeply shaken. I was horrified by it. As *soon* as I saw what was going on, and had some sense of what was going on, I switched, *automatically*. I became vegetarian *instantly*. And it wasn't a struggle; it wasn't difficult. I just couldn't even believe what I came to understand. But it's hard to get a full picture of what's going on, especially when you're young. I only had a very partial grasp of the animal agricultural system. So I started out as being a vegetarian and as I started to look further, I realized that the same kinds of processes are at work with dairy, and eggs, and other aspects of animal agriculture. It was hard to piece those together. For people who become vegan and vegetarian today, there are *so* many resources: There are books, there's YouTube. When I was doing this in the mid-1980s, I would say that there was very little information. I pieced things together bit by bit, looking at activist literature and reading popular books and scholarly books, so as that picture started to form and I got a full sense of what was going on, I became vegan fairly quickly. In both instances, as I reflect upon it, it's very strange. It's almost as though the decisions are made before you even make the decision.

SK: What are your thoughts on the term *vegan* as an identity or a category of analysis?

MC: In terms of the vegetarian and vegan divide, I would definitely make a distinction between the two, especially when I'm talking to people who are aware of the distinction. When you use the term *vegetarian* as it usually circulates, it refers to people who avoid eating meat, plain and simple. It might be for ethical reasons, having to do with animal welfare, or it might be for health and nutrition reasons, which are popular reasons these days. Some people go that route for environmental reasons, but, whatever the reason might be, vegetarianism is usually limited just to "I avoid eating meat" and it stops there. And I don't have any issue with that, and I'm not trying to be critical of that, just as a heuristic, an analytical distinction, that's how vegetarianism is usually understood.

Veganism, at least in my experience, seems to stretch and keep going quite a bit further. In popular terms, veganism is avoiding eating meat, as well as avoiding eating dairy and eggs. But then you start raising questions about leather, and other byproducts. Veganism has this tendency to ask questions related to harm to animals, and it tends to spread. I don't know if this is the right word, but I always think about it in a kind of obscure, strange logic of contiguity, where you start with one question and it abuts to another one, and that links up to yet another one. Being a vegan makes you constantly keep asking these questions.

I think that's where the impossibility of purity arises. For vegans it seems that that logic of contiguity pushes you to ask a series of questions and pushes you to probe further, and the more you probe, the less pure you find yourself to be. The logic of contiguity, or the kind of "spreading of concern" that you see among vegans, pushes you to ask questions about eating practices and consumption practices, and the ways that animal products are found in places you wouldn't expect to find them. So if you talk to a vegan, almost all of them have their own version of "These are the ten most surprising 'vegan' foods that aren't actually vegan." So by pushing those kinds of questions, you find all the ways in which animals are used, and in which violence against animals pervades and saturates the social sphere. Pretty quickly you discern that purity is not going to be possible. And that has implications for how you think about ethics and politics.

The other portion of that spreading of concern and that contiguity way of thinking is that you start raising questions about what *consumption* means, and how that term spreads, and all the ways in which we consume animals, whether it be literally or symbolically, and all the different kinds of uses we put them to, and all the ways we think about them violently, conceptualize them violently, and also, in the more literal sense, the indirect harms and direct harms we're caught up in and that we don't think about unless we push ourselves.

We're working right now on a project on automobility called *Alter-Mobilities*. And it's looking at the various ways our mobility system is caught up in doing violence, not just to animal life, but to ecological life, and all different aspects of other beings with whom we share the planet. So as a vegan, you might start asking questions about how you move, how you transport yourself, and the practices that one is caught up in there. So veganism for me is a constant pushing of those concerns. So that's why I would identify myself with veganism. Vegetarianism tends to stop the questioning at a certain point, and veganism insists that it remain open and keep carrying on.

SK: I am struck by your thinking here because it runs counter to the prevailing notion of veganism as a closed, puritanical approach to the world. Is there anything you want to add about how your thoughts and feelings regarding the consumption of animals have changed over the course of your career?

MC: My ideas about veganism have changed over time and have deepened over time. I will say from the outset, though, that I never really thought that it was something about which you could achieve theoretical or practical closure. Once I got into that space and realized what I was caught up in, it was an abyssal kind of thing; it changes pretty much everything about how we think and how we live.

SK: Could you describe the relationship between your approach to philosophy, your writing, and your veganism?

MC: I'm strange, with regard to philosophy, at least on the contemporary scene. I think philosophy as a professional discipline is bizarre. I do *not* see myself as a professional academic. I think philosophy is a way of life. Pierre Hadot is a scholar of ancient philosophy. And he shows that what we call philosophy today is very different from how the ancients practiced it, and I'd say in this sense I'm much closer to the ancients, in that philosophy for them was an entire way of life. It's an existential *commitment*. It's not something you do in the classroom or on the side; it saturates and pervades every aspect of your life. It's a way of calling into question the natural attitude that we have. We could call that natural attitude "anthropocentrism." So veganism, for me, is an ongoing set of rituals, and practices, that call into question that natural attitude; it's an attempt to hold it at bay, to give yourself some distance from it so that you can reflect upon it, because we're inserted into a set of formative practices concerning meat eating that we *never* reflect upon, a set of formative practices having to do with the use of animal bodies, and violence toward animals, that we very rarely think about. Veganism lets you see that system, reflect upon that system, distance yourself from that system, and ultimately transform that system.

Likewise, writing is an *essential* component of doing philosophy. Writing is an attempt to think through the kind of limits that are encountered in practice, to some extent. So as an activist, or as someone who's engaged in animal issues, you have your nose to the grindstone, you're working hard on campaigns, you're working hard on a particular issue, and you rarely reflect upon how this strategy might fit into a larger framework that would be critical. So writing also allows you to do this distancing thing where you're pulling yourself away from your natural attitude, but you're also pulling yourself away from your vegan practices and reflecting upon those as well.

Likewise, being a vegan reverberates back on the writing. A lot of the time, when I'm writing, I write myself into dead ends. I have no idea what I'm doing half the time. But it's in *practice* that I find some kind of answer to those dead ends, or some way to push through. So practices create passions, perspectives, and interests that push writing along, that help address limits inside writing. So there's a strange dialectical and counter-dialectical relationship, between writing and veganism for me, and they both play an essential role in philosophy as a way of life and as an existential commitment to live differently.

SK: At the end of *Thinking Through Animals*, you write about Donna Haraway's work and how it has shaped your thinking, but you also distinguish your approach from hers. Whereas Haraway argues that our relationship with domesticated animals as laborers, subjects of research, and companions should be reworked, you suggest that such relationships should be eliminated as much as possible. Can you elaborate on your perspective on companion animals?

MC: Of all the questions you've asked me, this one's probably the most difficult for me to answer, in the sense that it would take me a lifetime to get an answer out. My relationships with animal companions, neighbors, and friends—to do justice to them would take several lifetimes. But the first thing I would say is that thinking about animal issues and spending a lot of time in this field has made me ask questions about the terminology, so the terms that I use—*companions* or *friends* or *neighbors* or *commensals*—all of those terms are meant to indicate that the kind of human–pet relationships or owner–pet relationships that we talk about would have to be completely rethought. Those other terms point us to a very different economy and logic of relations. The underlying ontology that's at work with the concept of pets and domestication also has to be rethought in a fundamental way. I don't want to downplay the fact that "pet relationships," as we call them, have an obviously violent and deeply uneven nature in their history. These are the sites of some of the most horrific and problematic forms of violence done to animals. But when we look at these, the standard discourses around pets and companion animals tend to see animals as lacking agency and give very little role to their ability to resist, to their passions, to their desires. So with regard to Donna Haraway, I'm very much on board with insisting that animals have a certain kind of agency, and that you can't think of domestication as a one-way relationship. It's very much a co-constitutive and mutual relationship. But the accents that are placed on *where* the violence is, and which kinds of relationships might be saved, and which ones might not—those are very different between us.

I'm deeply critical of the vast majority of what goes on under the name of pet ownership. And when those relationships with pets take place inside the machinations of market logic, they get accelerated and intensified in the most violent ways that you can imagine. So if we're going to talk about commensal animals, we have to start thinking about those relationships outside the kind of reductive economic circuits that dominate our thinking. I don't have an eliminationist stance on companion animals *per se*; I don't think that every kind of companion relationship needs to be abolished. I know there are some vegans who are like, "Oh, we've gotta get rid of dogs, we've gotta get rid of cats, we've gotta get rid of domesticated animals altogether." There is a small segment, I think, of vegan and animal rights activists who make those statements, but I don't think doing so is characteristic of most. It's too late for that anyway, you know. There are too many relationships, and too much water has gone under that bridge. These relationships precede and exceed us, so it's a matter only of *what kinds* of relationships should we have going forward.

SK: You've written about indistinction in multiple venues, and in a variety of ways, but I'm wondering if you can review that idea for us here.

MC: I typically relate indistinction to identity and difference. So "identity" is the position that dominates inside of analytic, or mainstream, animal philosophy, and that's the position associated with Peter Singer, Tom Regan, those kinds of thinkers. And what they're doing is trying to show some kind of similarity or identity between humans and animals. And if you can show that identity, then whatever consideration, or political or legal standing humans have, should be given to animals on the basis of "They're just like us," "They're identical to us," or "They're relevantly similar to us." I've never felt any affinity with that kind of approach, to be honest with you. I don't want to denigrate that work; I think it's extremely important, and it raises all kinds of questions, but it's just not my particular approach.

While I was reading that material and not being particularly pleased with it, I was also reading Derrida and other continental philosophers. That resonated with me a bit more, and also the feminist approach that's associated with Carol Adams, Josephine Donovan, and Marti Kheel. There was something going on there that I thought was getting closer to how we might think differently about animals. So the feminist theorists would focus on care, and relationships; that seemed important to me. But, like most people who are my age—I'm in my early forties—I came up with a different brand of feminism, and a different set of ontological, ethical, and political commitments. So while that work resonated with me, I wasn't really situated in it fully and didn't accept all of the analysis.

With regard to Derrida, I was close to his idea that you could extend respect, care, and passion for animals without trying to assimilate them to us, or trying to make them identical to human beings in some way. The entire philosophy, logic, and ethics of difference that flows out of Derrida and Levinas are very useful for thinking about animals. Tom Regan will say, "The reason we should care about animals is because they're subjects of a life, and they exhibit all kinds of characteristics of subjectivity." That's true in the case of some animals, but it's not true, probably, in the case of others. So what do we do with these other entities that don't fit our schemas? Derrida gives you some way of thinking about those beings that are left outside of those schemas. At the same time, his work has always struck me as being, on the one hand very open and very generous, but on the other extremely conservative. So with regard to the human–animal distinction, I've never felt that distinction did any work that was useful. For Derrida, it's a matter of "Oh, we can't get rid of the distinction, we have to refine it and complicate it, see how it folds on itself, and look at the way its edges bleed." And that's fine, but it also blocks you from seeing radically different kinds of differences, and radically different kinds of relationships. Indistinction, for me, was an attempt to pull away from the endless complicating of the human–animal distinction, or the attempt to make humans and animals identical. It was an attempt to create a broad space for reflection. It *wasn't* an attempt to say, "Here's the new framework that everybody should follow." That's, I think, the major temptation of everybody who works in philosophy: "Okay, so what's your ontology, what's your ethics, what's your politics, and let's get it down and you tell me what the whole new framework is." I definitely was *not* trying to do that.

What I do want to say, though, is there's got to be *some* delimitation. And for me, that is to say that anthropological difference no longer serves as the departure for thought. We're not concerned, first and foremost, with how human beings are separated or distinguished from animals. When you set that distinction aside it allows the differences that we're talking about to be distributed along other axes; for other identities, differences, relations, and assemblages to open up that you might not have seen otherwise.

SK: What are some of your hopes and ideas about how critical animal studies might continue to evolve, and how do you see non-Western discourses fitting into this evolution?

MC: There's any number of ways in which you could work this field over, and have monistic tendencies, and say "Let's find *the* framework; surely they can't all be right." That's the game philosophers want to play, and that's the game I want to avoid at all costs. I want to leave the field open for the moment, and I want to leave it open because I think there's a need for plural ontologies. And "plural ontologies" means *many* ontologies, not just a pluralistic ontology. And, also for someone who's inside my subject position—a white man inside a settler colonial society—it's important for me not to come inside this zone and play the role that most male theorists do in this zone: offering up their framework and trying to saturate and dominate the field.

The important thing is that critical animal studies not just revolve around itself, but that it open up to other discourses and other traditions. For me, the most interesting and most promising work in critical animal studies is at the intersections with critical disability studies, race issues, and gender, environmental, and climate justice struggles. The most inspiring sources of how to think and live differently with regard to animals and other forms of human and nonhuman life come, typically, from non-Western and Indigenous sources. My good friend Brianne Donaldson is an expert on Jainism; then there's the Bishnoi people, for example. Of particular importance to me—I think you're going to be talking to Kim TallBear?—is a point that she makes about how animal studies and multispecies ethnography has a lot to learn from ongoing Indigenous traditions; and ongoing Indigenous discussions about these issues, I think, is *extremely* important.

It would behoove those of us in animal studies to stop for a moment, to pause, and engage in a humble and generous dialogue with these traditions and think more carefully about them. For my own part, this is what I've been doing for twenty-five-odd years. Those discourses are extremely important to me. I don't write on them for various reasons. But I think it's important for critical animal studies people to realize that there are *massive* insights, and longstanding practices and traditions, inside alternative cultures, and histories, that they can make use of. So with indistinction I'm trying to make room for those other experiments in thought and practice.

SK: You've demonstrated a clear commitment to bridging the rift between environmental and animal ethics, and while the claim that "Plants have feelings too" is often used ironically—to dismiss veganism out of hand—could we ask you to discuss how the concept of indistinction might extend to plant life? Or to micro-organismic and bacterial life?

MC: The linkage between animal and environmental issues is hugely important to me. I'm largely focused on how we can get critical animal studies to open up to the intersectional stuff, but also environmental issues. So yeah, as a vegan you get those questions all the time: "Don't plants have feelings too?" or "Isn't the carrot in your salad screaming out in pain?" And all that is intended to make vegans look ridiculous, inconsistent, or hypocritical. When they come from that angle, those questions aren't particularly interesting, but they *are* interesting when they come to you from people who care about these things and are sincere.

What *is* the response to somebody who asks in a kind of heartfelt and sincere way, "What about our obligations to plant life?" or to micro-organisms, or to soil, or whatever it might be? The one tendency among vegans that I want to avoid would be backtracking, and saying, "Well, plants aren't sentient" or "Micro-organisms aren't sentient," or "They aren't subjects of a life, and therefore doing violence to them doesn't matter." That approach is clearly not the route that I want to go. Everything I do is trying to avoid creating those zones of sacrifice, those zones of violence, where you can kill with impunity. So when we eat, I think questions should always be raised that are troubling, that produce anxiety—to use Chad Lavin's term—or to use your term, "messy." Whenever you eat there are messy questions that surround it, and you can't stop that messiness. So I use indistinction in view of that kind of approach. My aim is always to find ideas and concepts that place human beings within conditions that precede and exceed them, that *aren't* proper to human beings. When I'm working on animal issues, I'll focus primarily on the ways in which the concept of indistinction might show how human beings might find themselves within or alongside animal life in surprising ways, but I don't want to stop there. I would also want to show how humans and animals themselves are situated within another set of relations that are surprising, and that participate in things that exceed *them*. There's no logic of propriety there; there are only logics of ever-more surprising and profound relations.

In discussing meat, I actually prefer to speak of "flesh." A concept like flesh should spread quite a bit: We think about "human flesh," we think about "animal flesh," "plant flesh," or "fruit flesh." That concept can be stretched a long way. So if we think about "flesh," there's a way of thinking about veganism, which is to say, "Human beings aren't made of flesh, we aren't made of meat, we're sacred inviolable subjects, we're fundamentally inedible. And being vegan is about taking animals, who are fundamentally like us, and moving them over into this sacred, inviolable, inedible sphere." I, of course, am moved by the people who do it the other way. The people who drag human beings into the edible sphere, the people who drag human beings out of the sacred sphere and put them back into a common zone, put them back into the flows of life, and so on.

So I would put humans, animals, and plants all in the edible category, and the question then becomes how to eat respectfully and also, how to *be eaten* respectfully. With regard to micro-organisms, bacteria, and so on—same thing. There's a website that's very popular among philosophers called *Daily Nous*, and they just posted an article that was highlighting all the different ways that what we call an individual human being is being eaten from the outside and eaten from the inside, constituted from the outside and constituted from the inside, by so many different kinds of organisms, such that it would make something like "an individual" impossible to even talk about. That's more the kind of perspective I would adopt. That certainly makes questions like eating, *messy*, and irreducibly messy. But that's fine, by me. You have to stick inside that messiness; there's no way to avoid it.

SK: You said earlier that you might have a little more to say about our sense that there seems to be a separation between critical animal studies and food studies. Can you comment on that?

MC: I would say that the general statement you made about there being a division is definitely true, with the exception of minor trends. Things are changing at the moment, as we're speaking, so I think your volume is timely. One of the blockages that have been responsible for the division between veganism and other issues in food politics is that single-issue focus that's been characteristic of mainstream veganism, where you can deal with veganism as an isolated issue and divorce it from larger social issues, economic issues, and political issues. That's a real problem, but that's changing. What's promising today is the number of scholars who think about vegan issues and critical animal studies issues alongside food justice and food politics, and they're dealing with everything from locavorism, to sustainability, to distributive justice issues, food sovereignty, trade, governance, and so on. So there are great folks, you know: Vasile Stanescu, my friend Richard Twine, Breeze Harper, Tony Weis, David Nibert. And then, almost all the scholars at the Institute for Critical

Animal Studies, they've made these issues central to ongoing work in the field. So that relationship is becoming more established. That's an encouraging trend, and it's only going to be accelerated and intensified going forward.

SC: In your work you have offered examples of how animals resist subordination. Can you discuss how you read resistance in animals non-anthropocentrically? Can you think of ways that humans are subordinate to animals and how these might be useful for thinking about or acting differently in our relationships with animals?

MC: Anybody who's an activist with regard to animals knows these stories. You know how widespread and fierce animal resistance is. The odd thing, though, is that if you're not an activist, or you don't actively seek this out, almost all of it's hidden from view. So, Jason Hribal's book *Fear of the Animal Planet*—it does a great service. It brings together all this information about acts of resistance from animals, that take place in zoos, slaughterhouses, marine parks, and circuses. What his book and the other evidence suggest is that we might want to think of resistance as ontologically primary among animals. It's not something that happens after the fact; they, on their own, push back. They bite back, they insist, they take stands of all kinds.

We didn't talk about Levinas, but one of the ways you could pull Levinas into this is through his notion of "expression." Expression is the way in which the face expresses vulnerability, subjectivity, interiority, even resistance. There's a way in which animals express themselves and don't need us on the scene in order to come to presence. Levinas insists, at least with regard to other human beings, they bring their own presence with them. They are *kath'auto*: They present themselves. Likewise with animals, they express *them*selves, they present *them*selves. This is important to tease out, for me anyway, because I have a real problem with the way in which some animal activism is presented as "We are a voice for the voiceless." I very much dislike that kind of discourse because it denies animals the ability to express themselves, and to have their own forms of presence. I would rather try to frame animal activism as a *response* to that ontologically primary form of resistance; it's a kind of support. We are coming onto the scene after those acts of resistance, responding to those acts of resistance, and trying to support them in various kinds of ways.

Now, for anybody who's trained in continental philosophy, you have to be very careful with the idea that you could read those acts of resistance properly. They're not transparent. We don't fully understand everything we see. There's always a filtering, an interpretive act. In terms of indistinction, the resistance that we find among human beings—it isn't just ours. It's something that we participate in; it precedes us. Forms of resistance are basic to life and death, *per se*; you find them, pervasive, throughout the universe. That's important for us to remember, that we are participating in resistance that exceeds us historically speaking, and exceeds us in the present; it exceeds our orbit. That's how I would frame the issue of resistance.

CHAPTER 13

Taking Things Back, Piece by Piece

Sharon Holland

Sharon Holland, a professor of American studies, describes food as a place where community and laughter come together. Here, Holland recounts her lifelong love of cooking as well as her love of theory, which she likes to work around, "like a really good wine, in the mouth." A longtime equestrian, Holland says she has to come to terms with her status over the horse and that this relationship, like her relationship to meat, is contested, negotiated, and constantly reworked. Speaking about her beloved dog Sula, Holland underscores this idea of dynamic relation saying that their relationship of care was both mutual and co-constituted. Last, Holland shares her thoughts on blackness within animal studies, offering that she would like to rethink human–animal relations along novel axes. We interviewed Holland by Skype on February 3, 2017.

SHARON HOLLAND: I was born in Washington, D.C., after a very raucous party on January 1, 1964. My dad was a med student; my mother was working at the Library of Congress. She was one of the first black women to graduate with a master's in Library Science from

UNC Chapel Hill, so I'm a tar heel through and through. My mother was perfect during her pregnancy: no bad food, three meals a day, no alcohol, she never smoked. She never drank very much—two glasses of champagne and she was two sheets to the wind. My birthing starts with food, or actually with alcohol. The story goes that my father gave my mother—imagine someone actually smaller than I am, very slight frame—he gave her a gin and tonic and said, "Drink this." I was supposed to be due January 14. He said, "Look, she's got everything she needs, just drink it and relax." I was probably *in utero*, just like, "La-la-la—oh my god, what is that feeling in my hand? I can't see it." It was a party across multiple units in an apartment complex inhabited by Howard University med students and their families. The story goes that my mother went into labor and told a lot of people, "Oh god, my water broke." There was a rush to the hospital. My dad was in a whole other part of the complex playing cards or something; they came up to him and they're like, "Carolyn is going to the hospital!" and he goes, "I'm playing cards!" And they're like, "No, seriously, she's having the baby!" So they rush my mother to the hospital, and of course she passes out, because she's really not a drinker. And I start to come and the whole party moved to the hospital. Somewhere in there I was born; my dad delivered me because they couldn't find the doctor who was on call. I was born in segregation—1964, before the Civil Rights Act—so I was born in a predominantly black hospital. My mother finally woke up when the nurse was at her elbow saying, "You really need to feed your baby," and she said, "*What* are you talking about?" Apparently, I was a really good latcher, no problems there, ferocious eater.

I spent my summers as a child in Durham, North Carolina, with my grandmother. So I consider myself a Mason-Dixon–line person, you know—between chocolate city and the segregated South. My grandmother was my foodie friend; my mother couldn't cook worth a damn. She gave me a Betty Crocker cookbook for my twelfth birthday. I asked her about it later; I said, "How did you know I would love to cook?" She said, "I had to do something, or we would have starved to death." She was a single parent; she divorced my dad when I was seven. I went from one part of Betty Crocker to the next. I'm sort of OCD, so instead of organizing meals I would just make things; sometimes there was pie for dinner, because that was the section of the book I was working on. By the time I was fifteen, I was pretty much cooking all of our meals. My mom would shop and I would be like, "No, I told you to get this cut, I need this cut." She would say, "Well, Sharon, they don't have that at the Giant," and I'm like, "Well, go over to Snider's and find it." I was kind of toppy when it came to cooking.

SK: Were you the only child?

SH: Oh yeah, can't you tell? I wouldn't say I was spoiled, because my grandmother had already made that mistake with my mother. But I was very, very fortunate to be much loved. I realized once I got to college how protected I was physically, in my body and self. I think that had a lot to do with food and care; they all came together for my family. The women ruled the roost, very southern, very matriarchal, very Irish.

VICTORIA MILLIOUS: Can you tell us about your education?

SH: School was a nightmare. I mean for nerdy kids like me? I got kicked out of public school because I was considered "unruly." I finished my work early, so one day I peaced out and just left school. I literally walked out the door, looked both ways on the street and figured I could probably walk myself home. My mother was so angry; she took me to my dad's office and said, "We have to put her in private school, I found a private school for her, write me a cheque." I was private-schooled from then on—from the first or second grade. I found myself at this amazing place called Barrie Day School and Camp, which was run by the amazing Tim Seldin. It was filled with hippies: Imagine climbing a ladder to a loft and waking up your art teacher. I mean we probably rolled joints for them when we were younger. All these things that if you did them now, people would be like, [gasp]! Everyone who went to school in the '70s said we were raised by wolves. We were thrown into environments and had to fend for ourselves, so being a child was always like being part of a posse. I loved going to school there. It was Montessori, so if you decided to peace out it was no problem. If you were done with something you could go to another room. We had a real sense of freedom *and* community.

We had animals on the farm, and that's where I learned to ride. We had a couple of Great Danes who used to wander in and out of classrooms. One of my formative moments there involved this rooster that got all cocky and would attack you when you'd go to feed the chickens. All the kids started complaining about it, so Tim Seldin held this community meeting and we decided it was time to kill the rooster. Of course in Montessori you couldn't do anything without everyone's participating, so we had to participate in its death. Imagine an entire school, 100 kids, as they chop off the rooster's head and it runs around spreading blood. We're like, "This is horrible!" There's crying, and wailing, and holding onto teachers' pant legs. We decided we were going to be vegetarians. That lasted for about a week in the cafeteria when finally one kid said, "Enough of this, I want a hot dog!" There was the inner omnivore coming out and we decided, "Okay, we can have some meat."

I was at Barrie Day School and Camp until I couldn't be there any longer. When I graduated in sixth grade my mother put me in Catholic school—the horror of horrors. Concrete playground, what was that? I cried for a week and I was very lonely. I don't remember much about those middle school years. We were the generation that desegregated the schools; many people forget that. While our parents maybe marched in the streets, we were the "proof in the pudding," and that was really hard because we couldn't stray. We had a mission to prove to everybody we were capable. My mother never told me, but they didn't want to let me into that school. So she packed me up with a nice lunch; I think it was fried chicken wings because that was the only thing she could do really well. I don't know how because I never learned how to fry chicken—Martha Stewart taught me. My friend Alex always said, "I cannot believe you are the only black woman in America who does not know how to fry chicken," I said, "Shut up, I'm sure lots of us don't." He goes, "But you fry chicken with Martha Stewart!" I go, "Nobody has to know." I was not deeply southern in that traditional sense. So my mother packed me off with a good lunch like that was my school—but actually it was a trial. My mother said, "Test her at the end of the week; if she does not show up number one or two in her class you can kick her out." And I aced every test in a week and they were like, "I guess she *is* okay even though she's black." So my mother experienced a lot of that for me. My mother was a fierce advocate for not only my foodie health, but also my education. I remember they tried to kick me out of school one day because I had the wrong kind of socks on. And I told the principal, "Don't call my mother up here from work for some socks. I'll change them tomorrow, just don't bring my mother!" But she did, and my mother came with her little purse. I was sitting in the principal's office and she gave me a kiss and said, "Hey, baby, how are you?" I'm like, "I'm okay." She goes, "Sit right there." My mother had this way of talking to white folks: She'd get real close and lower her voice, and I heard her murmuring in there with Sister Marie Claire. When she came out she grabbed my hand and said, "C'mon, baby." I looked behind me and Sister Marie Claire's face was as red as a beet.

I guess my most formative moment was at this one school that I would eventually get kicked out of. My first day I was so excited, I had my little satchel and I was walking as usual not paying attention and I smashed into a pole and almost knocked myself out. The nurse was taking care of me in the kitchen—I don't know why—but the cook who was in there was an African American woman and she took one look at me and she sat me by the stove. The kitchen has always been kind of my refuge. That memory came back to me. Why was I always in the kitchen at school? I would just go in there. . . . I think the way we regulate children nowadays is a shame. Because of all the experiences that I had, although rolling joints for my teachers is probably not the best example. . . . But in public school I realize now that food was something I could experience because I could actually go into the kitchen. Now there are rules saying children should be in certain areas and line up in the hallways. So I'm really thankful for the '70s. That's how I came to food as a child. My grandmother was a very good cook. She taught me how to cook and had people who cooked for her; she came from a pretty bourgeois middle-class family and so I would go and get the pies and she had a cake maker and a roll maker. But I used to also cook at the stove from a very early age. With four, five pans going on, and this ginormous electric stove my grandmother had which was like a beast in the kitchen. I think she learned to cook because she was the oldest girl in a large family.

VM: What about university, your college experiences?

SH: I went to Princeton and the food there was, of course, not so good. There were these little kitchens that students could use, and I would get my own ingredients and make spaghetti dinners for my friends. People started coming to them and one time all these people were lined up and I was just like, "What is going on here? I don't have the money for this, go away!" And I was part of this vegetarian co-op at Princeton called the "Cat House," because it smelled like cat piss. There were all these feral cats running around, sometimes inside the house, sometimes outside. There was a bunch of us nerdy, English geeks. There was a guy named Byron there so we called him "Lord Byron" because we were all into Romantic poets; we thought we were going to save the world with poetry. But it's where I found my feminism and my love of literature and my first queer love. A friend and I were very much in love with one another; we both had boyfriends . . . but food, and the erotics of sexuality—everything was there in that Cat House kitchen.

Until the time that I ate the wrong brownie. . . . There were always two pans; one was loaded, one was not. Right before exams like an idiot I grabbed the brownie and went off to take my exam. That was a very interesting day. I was just like, "Crap, I can't focus on anything," and when I came back everyone was looking at me, and they were like, "Are you okay? Oh, it's Byron's fault! He left it out!" I'm like, "Wow, that is going to be the weirdest Chinese history exam ever." So there have been some missteps along the way. Like the time I put salt instead of sugar in a cherry pie. That was the grossest food experience of my young life. My mom was sweet; she actually tried to eat it, but then on her second spoonful, she said: "Oh, baby, I can't do this."

I do remember that she loved pepper steak—I don't think they even offer it at Chinese restaurants much anymore. I went to the cook of the restaurant we liked and I said, "Can you tell me how to make some pepper steak for my mother?" And he gave me a little jar of spice and I took it home and put together the shredded beef and the peppers and made pepper steak for my mom for her birthday one year. And she cried. My mom—though a formidable person—was not always emotionally present. I remember those moments where I was like, "Whoa, you're crying. Cool, I can make people cry with food, this is an astounding power."

VM: So how did you become an academic? And how did your interest in feminist, queer, and critical race theories emerge through graduate work?

SH: As an undergraduate, I was part of the women's center at Princeton, and there was a lot of stuff going on about rape culture on campus. We were organizing Take Back the Night marches and trying to get more representation from minority students on campus. I was also involved in the sanctuary movement on campus. I guess my first home was feminism. I met Lila Karp, who is now deceased; she was one of the original feminists from The Feminists in New York with Anne Koedt and Ti-Grace Atkinson. She gave me Celestine Ware's book. Ware was actually the first person to write a full-length treatment of feminism and she is a black woman. A lot of people don't know that; they start with Kate Millett's book. So I was led toward feminism by the first feminists, so I consider myself not a biological inheritor but definitely an intellectual inheritor of the discipline. Which really helped me because I knew black experience, but what I was experiencing on campus as a girl-into-girl-woman-child was ridiculous. People were getting sexually assaulted and they were distributing tapes of the sexual assault as "fun" in the men's clubs. Once I went to a Take Back the Night March, and the men had been gathering all day on Prospect Street—which is like the fraternity row at Princeton—waiting for us to march down. They were drunk and my then-boyfriend, who is still a dear friend, was a videographer at the time. So he put together a video of our night. He went up to men with his camera and said: "I'm taping you, I want your consent, what's your name? And why are you here?" These men-guys were so drunk they were like, "My name is John and F-U-C-K!" It was crazy and embarrassing, and also pretty astounding. He made three copies of the tape. He put one in the vault we had on us; we sent one up to New York—someone literally drove it to New York to a friend of Lila's who was on ABC News, and we got coverage. And because we had the video and we had their names, we decided to take them to small claims court—all of the men who tried to verbally and physically assault us that night. Then after all of this came out, the Princeton officials got in touch with us and they were like, "We need the tape." And we said, "Do you think we're stupid enough to give you the tape so you can destroy it? And do you think if we gave you the tape we don't have several other copies? You need to sit down with us and talk." The next week we had this huge rally, over a thousand Princeton students came, it was solidarity over women's issues from people of color at the Third World Center and from students in the Sanctuary movement. It was amazing.

What I'm trying to say is what I experienced in college was complete coalition among people of color, people in immigrant populations, Latinx people, black people, women of all colors and nationalities in solidarity. They used to call me "*that* Sharon Holland." So I did a lot of feminist activism and then I went to graduate school and tried to remain under the radar. I decided, "I'm tired of fighting, I just want to just go to school." At the time UCAR, United Coalition Against Racism, was gearing up on campus. This guy used to always come up and try to talk to me. He would go, "Are you Sharon Holland? Did you do all that stuff at Princeton? Because you need to join us." And I would say, "No, I don't," and he would say, "Oh yes, you do." And that was the beginning of my work with UCAR.

As an undergraduate I had done work in feminism, and Lila Karp was the one who gave me Audra Lorde's *Zami*, and I was like, "Oh, I'm a lesbian! Okay, I get it." Then I wrote my thesis on black lesbians and black feminism. Part of it ended up being published right out of undergrad in

Critical Matrix, a feminist journal. Then I went to grad school at the University of Michigan, moved with that crowd, and it was through working with UCAR that I really got my critical race theory on. I came out in my first year of graduate school. That coming out was peppered with my dad's death; he died from a self-inflicted wound in '88. I matriculated into graduate school in '87, and that was the occasion for the subject of my first book. Looking back, the experience of my dad's passing really put me on the outside of all the communities I traveled in. People come up to you and say, "Oh, I'm so sorry," and they want to mourn for you and your experience. And you tell them truthfully, "My dad killed himself." And you can see them receding. So I felt like there was something beyond the experience of race and queerness and feminism that I was involved in, something intangible. I didn't know what that space was or how to define it until I encountered Michael Taussig's work late in graduate school.

My father had always said that he would get me a dog when I got a real job. I was like, "Come on, Dad, a real job? I'm never going to have a real job." He was a medical doctor and always thought I would take over his practice. But I called him when I was fourteen and told him, "I'm going to be a poet." He stuttered, "Well, if you're going to be a poet you need a g-g-good job to support that." I was like, "Okay, Dad, I got that." After he died I went and got this crazy Shepherd and named her Sula Mae Peace Beloved. Please don't judge me for that. I loved that dog because when my dad died I came out to my mother, and our relationship deteriorated and never really fully recovered. She was a black woman of a certain generation who imagined for herself a certain configuration for her child's life. The one I am in is not such a configuration. So when my dad died, because of my mother's power in that family and because of the nature of his death, not one person in my bio-family came to his funeral. I felt completely disconnected from that thing that is supposed to sustain you. Thank god for my grandmother, who had given me enough, by that time, of what blackness truly is, so I didn't suffer too much. But I got Sula—I went to the pound and looked at the dogs, and there was this one sitting outside and I'm like, "Oh, what's that?" And they said, "Oh no, lady, you don't want that dog. This is its second time here, it's going to be euthanized." And I said, "Can I at least see the dog?" They were like, "I don't care." So I opened the door to the crate and I crossed my legs on the floor and she came out and curled up. She died when she was nine—way too early . . . I'm going to cry, I still miss her. I didn't know until much later; I thought I was taking care of her, but I realized she was taking care of me. I took that dog everywhere in grad school, I scared Lem Johnson damn near to death because he did not like dogs at all. I'd get on the elevator, and here's a black woman with this sable-colored Shepherd who's mostly off lead. Every black person was like, "Sharon, do you know what those dogs are used for?" I'm like, "I'm rehabilitating it. This dog is being rehabilitated to *us*." My dad was a big dog lover; he had three Akitas and two Shepherds. The big thing was to call the dogs into the kitchen, then run upstairs as fast as you could and get in the bed so they wouldn't all be sleeping in your bed, or there would be nowhere to sleep. It was the antithesis of my mother who, when I wanted a dog, gave me a toy poodle. But I loved Sula and when she passed I got letters from all over the country. People truly mourned for her because she was part of the fabric of our community in Ann Arbor. She was at marches and rallies; she was always in the bed of my truck. She went back and forth when I was going to Wyoming, she came with me when I did my research for my book at Indigenous reservations; she was my family. I guess this animal book will have one dedication: To Sula, who saved my life.

Around the time of my father's death, I also took solace in the kitchen. To avoid writing my dissertation I got the ginormous copy of Lord Krishna's *Vegetarian Cuisine* and worked my way through Indian cooking. I used to have these big dinners where people would come at nine o'clock thinking that food would be ready and I would be like, "Oh, you need to soak that for an hour." And they would eat until two in the morning. I don't know what I was doing, but they were so much fun and people would stay in the house until dawn just eating, playing cards, and talking politics. I would make breads, and dals and stews. Sometimes I'd be like, "Well, this didn't turn out well. Put the salt on the table." That's where I got my love of Indian food. It's still one of my passions; I make my own spices and give them away as gifts now to friends. When I was an assistant professor at Stanford, my friends would say to me, "Sharon, my mother is coming into town, can you please make some sambar powder and some ground masala and give it to me? She's always on me about my cooking!" Food has always been a place where I could laugh and get people together. I realize now that there were a lot of queer people at those dinners and we were in the midst of the HIV/AIDS epidemic—many of us were in ACTUP or Lesbian Avengers or other POC-powered activist groups. We ate to live and to mourn because the world seemed to be falling away at our ankles.

During that time I became a vegan, and that didn't go so well. I can't process vegetable protein as well as some people, so I lost a ton of weight and my doctor was like, "Here are some iron tablets, but if you take them every day it will shred your liver by the time you're fifty." So I reintroduced meat protein into my diet. Most of the queer women in our community at the time were vegans or vegetarians. So I joined the vegetarian and vegan co-op when I was in graduate school; we shopped at the co-op before Whole Foods even knew it was Whole Foods. Organic cooking was horrible at that time; you would come home with a basket of organic food and be like, "Oh, this is so sad, so sad." That was the late '80s, and I stopped shopping at regular grocery stores around then. So my relationship with food has always been about ethics. My grandmother was a big ethicist in our family. You asked me what my political practice is, so it might be queer, critical race, and feminism—but it really is toward ethical action.

VM: What do you see as the primary purpose of your academic work; why do you do what you do?

SH: I love to write, I love teaching and engaging in ideas. I love theory, I love to work it around like a really good wine in the mouth. My primary practice is really critical: Just because it's out there and it's feminist doesn't mean it's always good. Just because it's out there and it's queer doesn't mean it's always good. Just because it's critical race doesn't mean it doesn't need to be critical of itself. I find there's either too much self-reflexivity or not enough. So my work really tries to engage scholars, teachers, friends, students, and community members, in ethical practice. My animal studies work helped me to come back to ethics in a profound way. I used an example with my students about children going to school hungry. I asked them, "What are your political bents? If you are a conservative, raise your hand." Then I asked, "How many people in here believe that children should eat when they go to school?" I said, "This is the one thing we agree upon, so why do we need politics to make that happen?" People are too bound to ideology. The liberals want more money for food programs at school and the conservatives want fewer social programs, but in the middle the kids aren't eating. Even if they do get food, have you seen what's on the lunch plates? It's ridiculous. That brand of politics does not take care; it's not an ethic of care. I try to move my students to think about what's dropped out between those two poles. I'm not saying that I'm really moving in the feminist ethics of care and traditions. But I'm coming from a black Atlantic ethical tradition, which is infused with ethical action through black vernaculars. When I was a child, an adult wouldn't wag their finger in your face and sit you down;

instead, they'd free- and indirect-discourse you to death. That is the black woman's soliloquy, made very famous by Toni Morrison's *The Bluest Eye*. It was a way to create accountability and community. It's very shaming to sit there one on one with someone and have them jaw on about your behavior. Better to catch it in *community* and say, "Some people in this house really think they got maids, but they don't have maids so some of these people need to come correct." And then everybody stops and does what they need to do, because everybody is culpable at that moment. I learned that through the black community but also in Tim Seldin's community approach to solving problems. We are all culpable. That's the principle of my work: to try and think about what we're not thinking about; to try to be accountable to the things we do that are awful.

SK: Can you talk about your work on the place of the animal in African Americanist discourse?

SH: One project always builds off another; I've been working on animal studies for the past ten years. I'd already decided when I first moved here with my ex that I wanted to get back to riding horses. When I got here I started to ride; the first time I engaged with a horse I'd forgotten how freaking big they are and I was terrified. The women there gave me a grooming bucket and put me in with this mare that just looked at me like, "Oh, this is going to be fun." I was like, "Oh shoot, don't kick the shit out of me, I need my brain." The barn owner was like, "Okay, you lift up the hooves to clean them." I'm like, "I'm not touching that!" Now it's so funny, I literally lift up my horse's hoof while I'm talking—no problem. Eleven-hundred-pound animal, right? People always say when you're a menopausal woman you need to do something that's challenging and I think riding is that for me—conquering fear, trying to have a relationship with a being you don't have a lot of control over.

One of the things that have intrigued me with the animal studies work is, You know those sculptures with the black jockey? I watched in the South, as I would visit my grandmother and over a discourse in African American work, how people were talking about whitening those statues—painting the faces white. And I was like, Those statues must have some kind of relevance. So I was on the Cape for vacation; I took the ferry with my ex over to Martha's Vineyard and we saw the play *Pure Confidence*, which is about the black jockey tradition, and I was astounded—I didn't know about this tradition. When I was a kid I wanted to be a jockey; I realized then that the desire to ride was not something that was antithetical to being black. The first sport in America was racing, and the first race was three enslaved beings on three quarter horses—all mares—I believe in 1773. I learned finally about the great jockeys: Isaac Murphy and James Winkfield, who was there during the Russian revolution and helped save 250 thoroughbreds from certain death. Crazy stuff like that. It was black men who brought the current jockey seat to England; they went over there and demonstrated it. The groomsmen, the jockeys, and even some of the trainers were all black; and until a certain point they were the winners of the Kentucky Derby. This new project is a way to return people not only to a history, but our *connection* to it. And it wasn't just that we were trained by white Englishmen to ride here: Africans were brought over from the continent who were very good with horses, to help train horses. So when I get on my quarter horse mare I'm connecting a line that goes all the way back to the seventeenth century. I'm not participating in a white bourgeois endeavor, and when people try to tell me that it is, I remind them, "You got kids, right? How much do you pay for day care per week?" I'm like, "Mhm, having a horse is much cheaper in the long run." Of course this was before I had kids; now I have kids and I'm like, "Oh god, this is killing me."

So this work is a way to remind people that I'm not doing anything that's an anomaly. It's a tradition that I'm helping keep alive. And it's not the Western tradition; I'm doing the English tradition, which has its own history with blackness, and there are a fair amount of black women involved with the sport. There's a book by a black female jockey, Sylvia Harris, who is out about being bipolar. She's written this hilarious book about how horses saved her. She writes this beautiful moving piece: "When I go to ride a horse I go and I *breathe* him, just breathe him in, and I exchange breath, and then we ride together." It's true, that moment of connection when the animal under you carries you over the fence is . . . exhilarating, and life giving. And it *is* a black tradition. So many things have been stolen from us, and I feel like we're taking some back piece by piece. There's tons of stuff now on black female equestrians online, lots of black riders connecting about their love of horses. What's not to love? Have you ever felt a horse's muzzle? Softest thing ever. And quite truthfully, when we'd make the horses' mash at Barrie Day School and Camp I'd eat some as well.

SK: How was it?

SH: Really good. Warm and comforting. I still remember that smell. I guess riding horses is messy eating, in many ways. Because I have to deal with an entirely, very white world. The first barn I was in was a really racist place. The second one had a sign in the tack room that said "Unattended children will be sold as slaves." The second time I came I noticed it. I rode as much as I could in that barn with that sign, until after Ferguson. Then I told the manager, and she goes, "I don't really understand the stuff you say about blackness sometimes, about what it's like to be black, in the wake of Ferguson, I'm trying to. . . ." This is a vegetarian, committed animal lover, very decent person. I pointed to that sign and said, "I have to come in here every week and see that sign, and now that I have children it's unbearable to me." The next day I showed up and it was covered up—not taken down, but covered up. And that sign can only mean one thing. So it's messy. I never said anything to the barn owner there because I wanted to ride, but once I had children in my wake, riding became too painful and conflicted for me.

SK: Could you discuss the place of blackness—or perhaps the absence of black analyses—within animal studies literature, and how your work might be contributing to changing that?

SH: First, throughout history black people have always had an intimate relationship with animals. Frederick Douglass writes about it, writes about horses actually. So I'm not trying to *create* a connection; I'm trying to demonstrate the connection that is already there. Most people, when they think of blackness and animals, think of Michael Vick and his dogs. And some parts of my book talk about that case. But one of the things that came out of that case that made me so infuriated was the idea that "Because of slavery, this is the reason why black people in depressed areas treat dogs in a certain way." That is the most ridiculous thing I have ever heard! One of the things I'm trying to do is understand the relationship between blackness and animal by rethinking and imagining what the nonhuman animal truly *is*. The way to get out of the problem created by the matrix, human.animal.blackness is to actually think about the animal outside the boundaries of descent. If we *don't* think of devolution, we don't think of the human's relationship to animals in terms of evolving into something other. To become animal is not to *descend* into anything; I try to work the human–animal divide in animal studies, and then articulate that relationship through blackness.

One of the most profound moments I've had on this subject was at a presidential lecture I gave in Washington state. Vocabularies of Vulnerability was the title of the talk. One of the things I wanted to do was move against the critics working in bio-power around issues of precarity, especially in regard to the constant precarity of black life. I'm not a black pessimist. There's a great joy . . . I mean, those enslaved men who rode—yes, they were slaves. But they were also doing something that they loved. So I was giving this talk and I was talking about devolution, and this was right in the wake of a police officer's beating—let's call it what it was—of a black female child in the South Carolina school system. And one of the first questions came from the chair of African American studies. He said, "People are saying online that if that was a puppy there would be all this outrage, but since it was a black child. . . ." At first I thought about lying, and saying, "You're absolutely right," and just leaving it there. But I decided to be braver and said, "I'm going to tell you the truth: We don't care about puppies either. Puppies are always pulled out as this entity that sits between us and the structures of whiteness and our relationship to them, and our relationship to the animal. But let me tell you, all those videos on YouTube about the beatings and the deaths of puppies are filmed by people who don't intercede on their behalf, who are more interested in capturing footage of the demise of the dog." Then I shared another story from a horse friend who has graduated vet school. She once spent a whole day crying because she went to her anatomy lab and by the back door was a pile of dead puppies; the instructor told her to grab a puppy and put it on the table and begin dissecting. We kill dogs en masse, we gas them, and they're used to advance the cause of medicine. I said to him, "We don't care about the puppy; we don't care about the little cute thing. It's just a prop that prevents us from having a real discussion about not only our relationship with one another, but how our treatment of the animal is an absolute reflection of the status of the human." That's the thing I'm trying to fight against.

SK: I'll just follow up on a couple things you said earlier. You mentioned at one point you became a vegan primarily because of the relationship that you were in. Could you tell us about that, and your current relationship to eating animal products?

SH: Now, for the most part I'm a farmer's market person. I like to shake the hand of the person who killed the cow. I eat meat sparingly; I tend not to cook with it at home. My partner was vegetarian before she met me. Now she has a number assigned to her as one of the vegetarians that the local burger joint has converted. I'm not saying that meat is necessary. It's a *source* of protein, it's a *kind* of protein. But I do believe in whole planet, and the inner omnivore. But like I said, I don't think I've bought anything in cellophane from a supermarket in a long time. Even if it's in the co-op here now at Weaver Street Market, I know the woman at Firsthand Foods, the names of the farmers who grow my food; I drive by those cows every day. I say to my children, "See those cows? That's the grass-fed beef we buy when you go to the farmer's market." I remember my nine-year-old saying, "Oh, great," and our youngest exclaiming, "That's disgusting!" I told them, "You can stop eating meat for a while if you want, until you come to terms with it." I feel that riding is the same thing for me: I want to have this relationship with horses, but I have to come to terms every day with my status over the horse—literally and figuratively. I feel like all relationships are contested and negotiated, and every day I'm reworking my relationship with meat. I would like to be political and say we can be totally vegan. I'm not sure that's possible. But as Americans, if the rest of the world had our meat habit we'd be out of water by now. It's a big problem in the States and that's why I try to eat locally single-sourced foods. I'm aware in North Carolina we kill 32,000 pigs a day, for the rest of the country and parts of the world. That is a great sadness to me, great sadness . . . pigs are extremely intelligent. . . .

SK: Do you want to tell us about your blog—why you write it, and where you get the inspiration for your posts?

SH: It's a blog, but it's also an anti-blog: You can see the last post is around a year old. I'm thinking of Moten and Harney's *Undercommons*—I try not to *belong* to anything and try to make my own way, and be as subversive as possible. The blog is a way to say that this idea that you have to comment on things in the cycle of news is ridiculous. Why can't I take a year to think? I'm *still* thinking about Ferguson and what I want to say. I think this sense of immediacy has moved us away from contemplation and care.

The blog is also a way to put together the three parts of myself: the writer, the cook, and the rider. Sometimes I write about animals; sometimes I write about cooking and animals. Sometimes a memory will come to me and it will just flow. I've written something recently about the 94 percent, about the recent election in the States. It's a really really hard time for my family and this country and I'm trying to figure out a way to write about it. And there's food in there, and there are animals in there. Whenever we start talking about the human we get away from talking about the animal. In animal studies a lot of people don't even interact with animals all that much. It's all theory. So my work follows the feminist tradition of actually *engaging* with animals, as Vicki Hearne and Donna Haraway have done. And what better way to know other beings than to know a being that could quite literally kill you? The horse doesn't have to establish a relationship with you, like the dog, which has put its lot in with humanity. So the blog is a way for me to think in quotidian ways about the theorizing I do and the communities I serve, and sometimes do stuff that is political—to tell the bold truth. The blog really came out of a difficult juncture in my personal life. I was driving back from Cape Cod to North Carolina having just ended a twelve-year relationship and realizing that the person that I had believed myself to be was dying. The blog was this reemergence of self and connection with wider community, but differently. I had this experiment that I would write it and I wouldn't talk about gender or race or sexuality. It wouldn't be a "black blog" or a "feminist blog." I felt the blogosphere is so tied to identities. "I'm writing from the perspective of this" or "I am in the pool of queer bloggers."

VM: Last question: If you had to name one key concept that guides your thinking, what would it be?

SH: Ethical action. My other friend would call it "taking names." We were in the mountains with some dear friends from Chicago, who are all still my foodie friends. And one of them said, "I'm teaching your book [*The Erotic Life of Racism*], girl, in my graduate class, and we read it out loud for the excellent call outs." I'm like, "Oh god!" But at a certain point I realized I couldn't write about life with racism without talking about my community. If we're going to put the scholarship out there, then people need to be ready to have that matter. I felt like something needed to be said about the loss of the black lesbian body. I wanted to theorize about that loss. Not "How we can get her back?" but "How can we leave her alone?" It also came from a place of desegregation; wherever I've been I've been one of the few visibly black, out lesbians on campus. That work of desegregation has been a real struggle, really hard and so painful and lonely. And I know there are a lot of academics who are all political and they say they're going to do that, and then behind the scenes, you get down to the faculty table and you're like, "Really? You write about these issues and you said nothing at that meeting?" There's always this disconnect: You might have a book that says you care about class and race and intersectionality, but your presence on this campus is toxic. I have experienced this toxicity and I'm reminded that academia is an entrepreneurial endeavor; sometimes it's not about putting beliefs into action, so I feel like my writing is a way to do that. The first thing I do is count my failures—when I say "we," I'm talking about myself too. I've looked away from "X" or done "Y" and I need to integrate that doing.

SK: One of the things I really enjoy about your work—both your scholarship and the blog—is that you're so present in it. But it's not a narcissistic presence; it's heartfelt and careful—a hard balance to strike.

SH: Thank you. I try to do that in my work because I think we have to come out about the fact that we care deeply about the things we write about. I try to work in a way that people can see my deep investment. I also try to bring out their emotions: Stop looking at it from this "objective view" and get in there. That's why I was excited about your project, *Messy Eating*. As my non-bio sister once said to me, "Life is messy." It's never one straight line.

Codas

Toward an Analytic of Agricultural Power

Kelly Struthers Montford

The interviews contained in this collection highlight not only the personal and political aspects that constitute food and eating but also the sometimes fraught and negotiated tensions between individual, community, and systemic factors. As the interviews show, in many different registers and contexts, eating is not merely a physiological necessity but is related to questions of structural access, ecological ramifications, and ethics. At the same time, how and what we eat is deeply intimate and affective. Eating inescapably thrusts us into relation with various others, human and more-than-human. In this volume, the interviewees are asked personal questions about their dietary choices in relation to their background and academic work. It follows that their answers are individual reflections in which they describe how they make decisions about whom and what to eat by considering the others with whom they are in daily or familiar relation, such as spouses, colleagues, students, and nonhuman animals. However, while many interviewees denounce factory farming, some position small-scale and local animal agriculture (i.e., "happy meat") as an ethical or improved alternative. Indeed, the promises of small-scale agriculture were a recurring theme throughout the interviews. This comparison, I suggest, is symptomatic of a general absence in animal studies and food studies—that of the specificity of the power relations specific to animal agriculture and the foundational private property status that constitutes and authorizes these relations. As such, I propose that rather than focus on different modes of producing animal flesh, we turn to an analytic of *agricultural power* that allows us to consider the relationship between private property, life, food, and ontology. I then propose a contextual ontological veganism centered on beingness and relations. Such a position, I suggest, is a potential method of resistance to settler colonial power relations that are shaping our ontologies of humans, animals, and land.

Sharon Holland, Kim TallBear, Harlan Weaver, and Kari Weil, for example, explain that they will buy free-range organically fed meat when possible, while being aware of the potential limits of animal products marketed as humane. Holland says that she likes "to shake the hand of the person who killed the cow." Kim TallBear explains that the human–animal species barrier that authorizes hierarchical thinking and relationships does not resonate with Indigenous understandings of life. For this reason, she explains that there is nothing inevitable about dominant food practices, including the supposed inedibility of humans: "I'm not saying you should go out and kill people for food, although then one could say you shouldn't go out and kill nonhumans for food either. But if we are capable of being the prey of nonhumans, I think ultimately, morally, we're capable of being eaten by each other." Despite this ontological argument, TallBear highlights the ways in which our food systems constrain choice and shape relations. Speaking about her consumption of meat labeled "free-range," she says, "These labels facilitate people's shopping choices. Do they facilitate a real change in agricultural practices? I do try to consume less meat and less things where I don't know anything about the history. But you're reduced to looking at labels."

Weaver also positions his eating habits as relational in light of the human labor used in animal agriculture. For example, he states that he tries to not support factory farming because of immigration raids occurring in these locations and instead opts for locally produced meat when possible. Weaver also aptly highlights that many of the local farmers in his area are white and have recently moved from elsewhere, transferring their wealth accumulated in urban areas to start these farms. As such, he points to the fact that these "back to the farm" practices that traffic, I would argue, in idyllic imagery are made possible "using inherited capital that comes out of centuries of oppression." For Weil, who identifies as a vegetarian, her purchasing and eating habits become *messy* when she is cooking for or eating with others. She explains that she will buy humanely certified or grass-fed products for her husband, who would instead likely purchase "some cheap brand" produced using factory-farming methods. For these authors, despite signaling the limits of these "ethical" labels, these products are positioned as an improvement over cruel, environmentally destructive, removed, and covert factory-farming methods. The issues highlighted in these interviews are those related to production methods and do not necessarily entail or catalyze a shift in what or *who* is being consumed. This in itself is symptomatic of the successful operation of agricultural power that ontologizes animals as always already food. More specifically, the successful operation of agricultural power constrains which kinds of questions we ask. As shown in these examples, the issue is not about the consumption of *farmed* animals but instead about the mode of production through which animal products are sourced.

Based on my analysis of the available literature, the contrast between industrial agriculture versus small-scale agriculture is a false opposition that focuses attention away from the operation of power relations that are the foundation of animal agriculture as an institution: private property. As I explicate below, an analytic of agricultural power confronts us with the property relations that traverse the farm, its labor practices, its ecological effects, and how the human and the animal become parsed through eating. As Lisa Guenther and James Stanescu argue, industrial animal agriculture relies on the de-animalization of animals and the deading of their life.[1] Namely, animals in agricultural contexts are denied their animality as embodied creatures that have their own interests and desires and that come to understand themselves through meaningful relations with others of their choosing. Instead, they are reduced to input–output machines that can be manipulated, arranged, and replaced depending on the needs of the business in question.[2] James Stanescu has argued that because the very point of the factory farm is to produce animal products, farmed animals exist as "deaded life": While alive they are imagined as the meat they will become when slaughtered, butchered, and served.[3] While Guenther and Stanescu aptly highlight the interplay between the de-subjectification and objectification of farmed animal life as a consequence of the industrialization of agriculture, I argue that these processes are not isolated to the modern factory farm. Instead, de-animalization and deading are distinctly related to the property status of nonhuman animals, which can be intervened upon and manipulated toward the ends of the humans and institutions that are their legal owners.

It is then my position that the problems associated with factory farming that are aptly highlighted by respondents in this collection cannot be remedied through a nostalgic return to small-scale, local animal agriculture. For example, as Holland explains in her interview in this volume, "I know the woman at Firsthand Foods, the names of the farmers who grow my food, I drive by those cows every day. I say to my children, 'See those cows? That's the grass-fed beef we buy when you go to the farmers market.'" This passage illustrates the relations shaping local food ethics: The consumer is in a closer relationship with the farmer, yet the animals remain deaded life, imagined while alive as nothing more than the products they will become when slaughtered. As it is currently practiced, small-scale and local farming might better be understood as highly successful marketing strategies premised on telling consumers that they are making ethical choices of which they can be proud. Pragmatically, however, consumers are supporting the very relations and often commensurate practices that they are told they have avoided through this choice.

Locavorism pivots on three tenets of ethical consumption: environmentalism, the support of local economies, and the moral treatment of animals who become or produce meat, milk, and eggs.[4] The promise of locally produced foods as better for the environment is tied up in a commonsense notion that the less distance an item is transported, the lower its ecological imprint. However, transportation emissions account for very little of food's overall ecological impact. The production stage accounts for approximately 83 percent of food-related greenhouse gas (GHG) emissions, whereas transportation from producer to retailer represents 4 percent.[5] However, not all "foods" are as resource intensive as others. For example, the United Nations Intergovernmental Panel on Climate Change found that "meat production contributes more greenhouse gases than the entire transportation industry, including all automobiles, combined."[6]

At the current rate of consumption (more than 55 billion land animals per year), more than 80 percent of the land currently used by humans is for meat production. This is the leading cause of species extinction, water scarcity, and biodiversity loss as rainforests and other areas are cleared to become pastures.[7] While promoted as an ecologically responsible alternative, small-scale animal agriculture requires more resources than factory farming. Free-range production uses more land and emits more methane and nitrous oxide than do industrial methods. This is because farmed animals are allowed (somewhat) more space, and because these animals take longer to reach slaughter-weight, as they typically have not been genetically altered in such a way as to promote rapid weight gain. Given that small-scale animal agriculture accounts for only 1 percent of the meat produced in the United States, it is hardly a large-scale solution.[8] Even popular "happy meat" farmer Catherine Friend reports that factory-farmed animal products make up 25 percent of her weekly diet.[9] There is simply not enough land, even when we include the potential transformation of land currently used for feed crops, to make this a feasible worldwide "alternative."[10] Diets that do not include animal products, however, have been shown to be a far more effective mode of mitigating climate change and ecological destruction: "[S]hifting from beef to vegetables for even a single day a week would in fact be more helpful in reducing greenhouse gas emissions than shifting the entirety of one's diet to exclusively locally produced sources."[11] The question of environmental impact is then not about local versus global production methods but whether the product in question is that of a farmed animal.[12]

Small-scale animal farmers also market their animals as living "happy" lives prior to their slaughter, but an analysis of routine practices carried out on local farms reveals otherwise. As Vasile Stanescu reports, the production of happy meat commonly entails the forcible insemination of female animals, unnecessary surgical procedures performed without anesthesia, as well as the long and traumatic transportation of the animals to (often) the same slaughterhouses where factory-farmed animals are killed.[13] In *The Compassionate Carnivore*, Catherine Friend recounts that routine practices on her farm prevent her from meeting animal welfare standards for her sheep.[14] Vasile Stanescu summarizes Friend's approach: "[S]he uses herd dogs, does not allow her sheep to lie down comfortably during transport, does not provide continuous access to shelter, and most importantly 'docks' (amputates) their tails without the use of anesthesia."[15] In another book, *Hit by a Farm*, Friend describes castrating baby male lambs without sedatives or pain relief. There is indisputable evidence that this practice causes excruciating pain in the short-term as well as chronic pain in the long-term.[16] Other practices that both factory farms and small-scale farms employ include "grinding up male chickens at birth; using animals who have been selectively bred into shapes which cause disease, suffering, and early death; forcing and repeating pregnancies; separating family members for profit; and killing the animals in the exact same slaughterhouses and identically 'inhumane' conditions as factory farms."[17] In their approach to farmed animals, the reality of small-scale farming practices reveals little practical or philosophical alternatives to factory farming.

I would further argue that small-scale faming is coterminous with the settler colonial and racial capitalist systems that sustain the settler nation states of Canada and the United States.[18] Small-scale animal agriculture cannot resist the property relations to land and to animals that are foundational to animal agriculture and that have been deliberately instituted as part of settler colonial projects in Canada and the United States.[19] Specifically, colonists imported the institution of animal agriculture, farmed animals, legal statuses of property, and ontologies of Western human superiority that structured ways of being and living in their homelands. Colonists positioned animal agriculture as the civilized manner in which to interact with animals, as well as the proper way to relate to land as a resource. English law also allowed colonial governments to make legal claims to land as they could show (as per their own legal metrics) that they had a productive relationship to the land in question.[20] Animal agriculture in these settings is then historically rooted in a settler colonial project of territorialization. The property relation to animals, as Lauren Corman aptly highlights in her interview in this volume, underpins and is reproduced through dominant eating practices. It is for this reason that I suggest it is more appropriate to consider food and eating as bound up in the exercise of agricultural power and its multiple expressions.[21] An analytic of agricultural power allows us to focus on its underlying logic, relations of property, and political deployment, regardless of its exact technique of production.

For decolonial, environmental, and ethical reasons, it is urgent that we develop a food politics that attends to and resists the settler ontologies of animal life and land as private property that are foundational to animal agriculture. Corman explains that for her, the importance of veganism lies in its "direct challenge to the property status of animals and the notion that they can be rendered as objects. It's kind of a daily personal boycott, a rich practice of eschewing the understanding of animals as objects or servants from the beginning of their lives to their deaths." One such possibility for a food politics premised explicitly on the recognition of animals as beings and not property is a contextual ontological veganism. Because ontologies are inescapably political, they are also contextual and can be otherwise.[22] It is therefore not the case that because animal-based foods are often thought of as the most authentic and "real" foods, these food ontologies are stable. Instead, how we understand the "what is" of food is the result of deliberate political projects and, as such, can be otherwise. It does remain useful, however, to continue to ontologize food, as doing so provides a framework for legal, ethical, and political action.

I argue for a contextual ontological veganism premised on a distinction between edibility and food. An edible product would be ontologized as food only pending an evaluation of whether the relations that went into the making of the product, as well as the relations that are produced by its consumption are ethically desirable.[23] This ontology recognizes that both humans and animals are technically edible, yet unequally targeted by agricultural power. In other words, the likelihood that humans will be subjected to the processes entailed in agriculture, from forced breeding and forcible confinement, to premature and exceptionally violent deaths, is minimal. This is because humans are persons under law while animals remain property.

Therefore, I propose that a true alternative ontology of food must also be premised on a non-anthropocentric legal ontology of animals. Maneesha Deckha's concept of beingness is helpful. *Beingness* refers to a legal subjectivity for animals in which they are neither property nor persons but beings that are relational, vulnerable, and embodied. Avowing the beingness of animals would remove them from the legal category of property in which they are reproduced as "living meat" and ontologized as food. An ontological framework such as this is responsive to the contextual realities and competing ways of knowing and being in the world that structure who is edible and in what ways. Such an ontological position recognizes that while we are all edible, we should not be subjected to the carcerality of animal agriculture. This ontology would then apply to our ethical evaluations about market-mediated food production and consumption, and is not meant to apply to nonproperty-mediated eating habits, such as the multitude of Indigenous subsistence hunting, gathering, or crop-based agricultural traditions.[24]

To return to the interviews that position large-scale and small-scale approaches as divergent practices that shore up different ethical practices and responses, we should instead understand such methods of production as different iterations of agricultural power that are produced by and that sustain the foundational relational of property. Within such a frame, it becomes apparent that there is very little that is alternative about small-scale animal agriculture. A focus on agricultural power might then be a necessary supplement to animal studies and food studies scholarship working to constitute ethical food politics. It is my hope that a relational and contextual ontological veganism can open possibilities for differently relating to ourselves, others, and to territory that has been not only violently dispossessed from its original stewards but also parceled into privately owned resources used to sustain animal agriculture and its subsidiary industries.[25]

NOTES

1. Lisa Guenther, *Solitary Confinement* (Minneapolis: University of Minnesota Press, 2013); James Stanescu, "Beyond Biopolitics: Animal Studies, Factory Farms, and the Advent of Deading Life," *PhaenEx* 8, no. 2 (2013): 135–160.

2. Guenther, *Solitary Confinement.*

3. Stanescu, "Beyond Biopolitics."

4. Chaone Mallory, "Locating Ecofeminism in Encounters with Food and Place," *Journal of Agricultural and Environmental Ethics* 26, no. 1 (2013): 171–189; Vasile Stanescu, "'Green' Eggs and Ham? The Myth of Sustainable Meat and the Danger of the Local," *Journal of Critical Animal Studies* 1, no. 11 (2010): 8–32; Vasile Stanescu, "Why 'Loving' Animals Is Not Enough: A Response to Kathy Rudy, Locavorism, and the Marketing of 'Humane' Meat," *The Journal of American Culture* 36, no. 2 (2013): 100–110.

5. Christopher L. Weber and H. Scott Matthews, "Food-Miles and the Relative Climate Impacts of Food Choices in the United States," *Environmental Science and Technology* 42, no. 10 (2008): 3508–3513.

6. Stanescu, "'Green' Eggs and Ham?" 13; Henning Steinfeld, Pierre Gerber, T. Wassenaar, V. Castel, Mauricio Rosales, and C. de Haan, "Livestock's Long Shadow," Rome: Food and Agriculture Organization of the United Nations, 2006, http://www.fao.org/docrep/010/a0701e/a0701e00.HTM.

7. Steinfeld et al, "Lifestock's Long Shadow"; Stanescu, "Why 'Loving' Animals Is Not Enough"; Stanescu, "'Green' Eggs and Ham?"

8. Stanescu, "Why 'Loving' Animals Is Not Enough."

9. Catherine Friend, *The Compassionate Carnivore: Or, How to Keep Animals Happy, Save Old MacDonald's Farm, Reduce Your Hoofprint, and Still Eat Meat* (New York: Da Capo, 2009), 240.

10. Stanescu, "Why 'Loving' Animals Is Not Enough."

11. Stanescu, "'Green' Eggs and Ham?" 13; Weber and Matthews, "Food-Miles."

12. Stanescu, "'Green' Eggs and Ham?"

13. Stanescu, "Why 'Loving' Animals Is Not Enough."

14. Friend, *The Compassionate Carnivore.*

15. Stanescu, "Why 'Loving' Animals Is Not Enough," 103.

16. Catherine Friend, *Hit by a Farm: How I Learned to Stop Worrying and Love the Barn*, first edition. New York: Da Capo, 2006); N. P. French and K. L. Morgan, "Neuromata in Docked Lambs' Tails," *Research in Veterinary Science* 52, no. 3 (1992): 389–90; Michael C. Morris, "Ethical Issues Associated with Sheep Fly Strike Research, Prevention, and Control," *Journal of Agricultural and Environmental Ethics* 13, nos. 3–4 (2000): 205–217.

17. Stanescu, "Why 'Loving' Animals Is Not Enough," 103.

18. Virginia Anderson, *Creatures of Empire: How Domestic Animals Transformed Early America* (New York: Oxford University Press, 2006); Billy-Ray Belcourt, "Animal Bodies, Colonial Subjects: (Re)Locating Animality in Decolonial Thought." Societies 5, no. 1 (December 24, 2014): 1–11. https://doi.org/10.3390/soc5010001; Claire Jean Kim, *Dangerous Crossings: Race, Species, and Nature in a Multicultural Age* (New York: Cambridge University Press, 2015); Kelly Struthers Montford, *Agricultural Power: Politicized Ontologies of Food, Life, and Law in Settler Colonial Spaces* (PhD dissertation, Edmonton: University of Alberta, 2017).

19. Anderson, *Creatures of Empire*; Kim, *Dangerous Crossings*; Struthers Montford, *Agricultural Power*.

20. Anderson, *Creatures of Empire*; Kim, *Dangerous Crossings*; Struthers Montford, *Agricultural Power*.

21. See further Taylor (2013) and Stanescu, where they suggest that we ought to begin thinking in terms of agricultural power. Chloë Taylor, "'Foucault and Critical Animal Studies: Genealogies of Agricultural Power,'" *Philosophy Compass* 8 (6) (2013): 539–551. https://doi.org/10.1111/phc3.12046; James Stanescu, "Beyond Biopolitics: Animal Studies, Factory Farms, and the Advent of Deading Life." *PhaenEx* 8 (2) (2013): 135–160.

22. Johanna Oksala, "Foucault's Politicization of Ontology," *Continental Philosophy Review* 43 (4) (2010): 445–466.

23. See Heldke (2012) on a relational food ontology. Lisa Heldke, "An Alternative Ontology of Food: Beyond Metaphysics," *Radical Philosophy Review* 15 (1) (2012): 67–88. https://doi.org/10.5840/radphilrev20121518.

24. Anderson, *Creatures of Empire*; Kim, *Dangerous Crossings*.

25. Belcourt, "Animal Bodies, Colonial Subjects"; Eli Clare, *Brilliant Imperfection: Grappling with Cure* (Durham, N.C.: Duke University Press, 2017).

Thinking Paradoxically

Billy-Ray Belcourt

In *Messy Eating*, scholars from a mixed bag of disciplines refuse the temptation of a diagnostic that enables those who, god-like, rummage through the muck of the social to hand down quick fixes for thorny ethical conundrums. Instead, the interviews collected here nod to a mode of attention that operationalizes rather than stomps out paradox, that does not shy away from the awkwardness of sitting with unanswered questions. We, academics and activists committed to engineering flourishing worlds, don't get to hover above the forms of dispersed suffering at which we aim a sociological eye. Sometimes we don't get to chime in. Eating, like all practices that produce uneven misery and killing, must be provincialized. With the interviews at our disposal, let's turn to the North, where paradox should guide how we think about situatedness, historicity, coloniality, animality, and the act of killing.

"Fuck PETA!" These were the sharply spun words of the Inuk musician Tanya Tagaq as she accepted the coveted Polaris Music Prize in 2014. Tagaq threw these words at the audience and out into the world, and the world is where they were picked up, turned over and over again, and looped through a settler structure of feeling that cruelly denies Indigenous peoples anything like a complex emotional life. In an interview with *The Globe and Mail*, Tagaq briefly nodded to the months of abuse she endured in the wake of her contentious comment. Dan Matthews, the vice president of People for the Ethical Treatment of Animals (PETA) at the time, was quoted in the same article; he too tossed about injurious words: "Tanya should stop posing her baby with a dead seal and read more."[1] Let us put the dead seal aside for now and pause at the social field that the declarative "read more" maps, for it is one that empowers murky public thought about Indigenous peoples and the coloniality of the present.

"Read more" is not a free-floating command; it echoes from the bloody maw of history. "Read more" coheres here only insofar as Tagaq is refused the quotidian capacity to pay quasi-anthropological attention to the world. Things are happening and Tagaq is a part of them, but Matthews makes her out to be a speaking subject whose words reverberate from neither context nor the social. This kind of utterance—that Tagaq "read more"—is part and parcel of an epistemology of whiteness whereby Indigenous peoples are imagined to be unaware of what one can do by way of the form and genre of political speech.

Matthews argues that PETA is not out to halt Inuit hunting practices, but rather that PETA's "fight" is against "the East Coast commercial slaughter," which he suggests is the wrongdoing of money-hungry white people.[2] Matthews wants to think of himself and PETA as outside of the shoddy social worlds in which Inuit are made to live and sometimes prematurely die, but what Tagaq registers via "Fuck PETA" is a political climate in which a so-called seal ban is put up for debate in the lawmaking theaters of countries throughout the Global North without much concern for Inuit livability. Tagaq's "Fuck PETA" is not necessarily about individual culpability but about the degree to which PETA is nested in a structure that atrophies Inuit food culture while also aiming public blame at Indigenous populations for social problems of the state's making. So, Tagaq ropes PETA into the matter and insists that the organization be held partly accountable for the damaged lifeworlds that Inuit inhabit in the wake of decades of biopolitical attempts to manage them as a population, to make them into a set of governable numbers.[3] This biopolitics of race is rhythmed too by a wildly accelerating state of global ecological collapse amplified by the clumsy interventions of southern scientists and activists alike.[4] "These protestors are taking food out of kids' mouths," Tagaq deplored.[5] She knows that the scale at which Inuit experience loss is worldly.

Now, what of the dead seal? This is where Matthews betrays his assertion that PETA has not sought to disturb Inuit ways of living. Indeed, Matthews is glib about the efficacy of Tagaq's social media practices, insinuating that her controversial "sealfie"—seal selfie—is of a piece with her supposedly "ill-informed rant."[6] In early 2014, Tagaq tweeted a photograph of her baby daughter posed with a newly killed seal. The sealfie was a "tongue-in-cheek protest" that sought to imagistically arrest the racial insensitivities of animal rights activists who choose not to attend to how they are caught up in the coloniality of the world.[7] It seemed, however, that Tagaq had represented that which was firmly outside what one might call, to paraphrase Veena Das, a public capacity to represent.[8] Her photograph was taken as evidence, for example, of her inability to mother her child; indeed, one Internet interlocutor, like Matthews, suggested that she lacked basic intelligence.[9] The seal was described with bloated words like *bludgeoned*, its dead body interpreted as a trace of what is lost when "Inuit tradition" is kept in the world.[10] But, the reaction to the sealfie sprouted into much more than a game of words—that is, Tagaq was sent digitally altered photographs of her and her baby's dismembered bodies. Here is how she described one of the disturbing photographs: "[H]e posted a picture where he Photoshopped my baby, there was a man stabbing my baby, and a kind of club-like thing across the front. . . . [T]hat's when I just kind of thought, ok, this is too much."[11]

How, we might ask, can someone oppose one image of killing with another? Does the second act of killing make up for the first? Following Judith Butler, is the grief so unbearable that one is incited to kill from that unbearability?[12] Does revenge surface here as a mode of political action, and does it perform a savage hatred for the proliferation of Inuit life in spite of the sovereign bloodlust of colonial lawmakers in the settler state of Canada? Is animal life so charismatically grievable that it stomps indigeneity into the abject gap between a past genocide and an ongoing social death? This is just one set of paradoxical questions we can ask about the digital abuse Tagaq encountered while trying to garner political attention in the public sphere.

I open with this vignette because it throws into relief a key thematic in the interviews collected in *Messy Eating*—that is, it begins to tell a thicker story about how paradox ensnares us everywhere, especially those of us committed to the project of engineering flourishing worlds for those who have historically been and continue to be made into surplus populations who must endure a brutish existence so that others can fantasize about a life lived without turbulence. In particular, I am interested in how Tagaq's sealfie compresses paradox into a single frame, how she goes about vocalizing an angered refusal of the murderous character of the settler state by way of an insistence on just animal killing.

Since at least the 1940s, the Inuit have been caught in the throes of a "biopolitics of care" that subjects them to the shallow sentiments of politicians who couldn't care less whether they live an animated life or not.[13] Colonial governmentality in the North constrains the modalities by which Inuit can participate in performances of worldmaking. Throughout the long twentieth century, Inuit were wrangled by government officials and other settler personnel to be sent to hospitals, sanitariums, and residential schools in the South, often to die. Their bodies were serialized, which is to say that the intimate lives of Inuit were stalled so that their kinship ties and entangled identities could be unearthed from an affective structure and replaced with names and numbers computable in the bureaucratic mazes of the federal government.[14]

Today, Inuit social worlds remain an unthought of sorts; in the imaginations of southerners, they exist "without history or politics," without complicated emotional entanglements.[15] Certainly, this was the diagnosis thought up by those at the Ellen DeGeneres show who described sealing as "one of the most atrocious and inhumane acts against animals allowed by any government," a description that partly roused the above-mentioned sealfie movement.[16] What we are dealing with, then, is a politics that offers restricted freedom, one that distills the subject of ethics by way of the hardened categories of race, class, and gender. In this day and age, might things be too paradoxical to hold on to the fantasy of total liberation? For activists like Matthews and DeGeneres, the ethical impasse was remedied quickly: The Inuit had to face the chopping block. The North, however, is where eating and ethics get messy.

"We are in and of the world, contaminated and affected." So goes Alexis Shotwell in her seminal book *Against Purity: Living Ethically in Compromised Times*, which takes as its object the charisma of a politics of purity in a disastrous geological time that scientists call the Anthropocene. In her words: "[T]he Anthropocene: roughly, the moment that humans worry that we have lost a natural state of purity or decide that purity is something we ought to pursue and defend. This ethos is the idea that we can access or recover a time and state before or without pollution, without impurity, before the fall from innocence."[17] Shotwell skillfully argues that to attach to an object like purity is to distract from the thorny ways in which we are enmeshed in "complex webs of suffering."[18] It is by owning up to our complicity in structures that harm some more than others that we can acclimatize to a politics that plants us firmly in the muck of the present, which is where we begin to dig an exit out of "the Mordor of industrial capitalism."[19] Shotwell comes up with a concept she calls "constitutive impurity," which in her assessment is how we, the impure and polluted, go about tending to a damaged world still worth bettering.[20]

Shotwell doesn't want us to run away from contradiction, so it is with her that I want to preface my thinking on paradox. Ours is a world full of contradiction and paradox. Paradox is descriptive of two or more scenes, encounters, or objects that butt up against one another for explanatory power. They compete for the ability to be thought of as symptomatic of social reality. At first glance, they seem to be dialectically opposed. But, paradox is a modality with which to tune into the cacophony of social life, with which to upend the hold that narrow ways of thinking have on our ethico-philosophical possibilities.

Everywhere we look, we see things that make us go cross-eyed. Everyday life is labyrinthine, so the more we find ourselves running into twists and turns, dead ends, and roundabouts, the less faith we instill in the categories that we inherit to make something of an ethical life. It is with these categories that we commit to a world that doesn't cage some in the badlands of modernity. But, at the same time, we give tacit consent to a polity, a settler state, that drags racialized populations through the mud of bad social structures and then into prisons to rot, physically and ontologically. We want a world without suffering, but we also have to grapple with the way in which animals underpin Indigenous life in the North.[21] Again, this is the key paradox with which this essay is concerned: how to make sense of the possibility that without animal death, Indigenous worlds would crumble?

If we think paradoxically, we have to give up on dogma. To think paradoxically, however, is not to think dialectically. The thesis-antithesis-synthesis triangulation that governs the dialectical is far too neat a conceptual framework for getting at the constraints under which life is precariously and sometimes extemporaneously slapped together in the North. Paradox weakens the epistemic power of normative theories that seek to once and for all determine how best to be in the world.

We can glimpse something of an anti-paradox stance in the purity politics of activists like Matthews and DeGeneres, those who let a desire for one type of flourishing obfuscate, for example, the tentacle-like connectedness of animals to Indigenous worlds. Without paradox, Indigenous peoples are squeezed into flattened forms of subjectivity with which we inch slowly and quietly toward a premature death. Paradox is how we refuse a settler–colonial optic that stomps Indigenous suffering into the rut of statistical truth. Matthews and DeGeneres turn a blind eye to the "entangled and complex situations" that make up the political in a place as semantically fraught as the north, where living and dying always hang in the balance.[22] Take, for example, Elizabeth Povinelli's claim that Indigenous social worlds exist in an economy of abandonment in which the lethality of state power manifests where we are trained not to see it: "Indigenous communities are often cruddy, corrosive, and uneventful. An agentless slow death characterizes their mode of lethality. Quiet deaths. Slow deaths. Rotting worlds. The everyday drifts toward death: one more drink, one more sore; a bad cold, bad food; a small pain in the chest."[23]

Indigeneity requires us to reckon with that which cannot be seen. It enables the not-seen to warp one's field of vision. So it is the not-seen of indigeneity, which is partly the not-seen of an encounter of slow violence, that paradox brings into focus. In a working paper called "Ethics as the Expression of Life as a Whole," Veena Das makes a case for a mode of anthropological talk that understands everyday life as the name for what is "both closest to us and the most distant from us."[24] Indigeneity is produced in the gap between "closest to us" and "most distant from us," and it is from this ditch that indigeneity shadows social thought. Indigeneity ropes us into a world that is polyphonic, and it is by way of this noisiness that we attune to an ethics that takes paradox as a condition of possibility. So, when DeGeneres calls the seal hunt an "inhumane practice" and when Matthews tells Tagaq to "read more," indigeneity spills from their mouths, falling flat onto the floor with their words.

It is paradox that allows us to negate what Kim TallBear describes as the ethical superiority that activists assume when attempting to steward "the lives of those that are less than they are." TallBear insists that these activists don't think about animals as relations, which is to say that they don't think of them as "simultaneously different entities," to use Todd's language, with slippery meaning that is always in flux.[25] To place animals like seals in a web of relations is to come to terms with the way that "there's no disentangling their lives from our lives now." It is from this inability to disentangle that we dis-identify from a politics of purity. Let's call this inability to disentangle paradox. It is by way of paradox then that we, to nod to Donna Haraway's key formulation, "stay with the trouble," which is the project of living in disturbed times, of settling "troubled waters" and making "quiet places."[26] But quiet places are nonetheless crowded with the loudness of paradox. Without the seal hunt, Inuit would hurdle toward even choppier waters. This is not pure speculation: in *The Huffington Post*,

Inuk filmmaker Alethea Arnaquq-Baril fixed attention on the aftermath of the 1982 European whitecoat sealskin ban: "There were all these factors that already oppressed the population so when anti-sealing came along and the hunters couldn't even feed their families, the suicide rate skyrocketed."[27]

Let's turn now to a passage in María Elena García's interview about the politics of killing: "I've been struggling recently with the idea of 'killing *as* care.' There's a famous case of a Bolivian woman who would rather kill her llama and sell her as meat to foreigners than sell her alive. So what does that mean in terms of thinking about those relationships, in terms of 'killing as care'? Is it possible? In that case, the decision was linked to spiritual ideas about what happens when you sell an animal alive without knowing what the consequences are, without knowing what will happen to this animal. I find it is important to think in a multi-layered way about killing rather than saying, 'killing is bad.'" Does this suggest, then, that sometimes killing is good? Does the act of killing ever emerge from an ethics of care? "Killing *as* care" is of course paradox writ large. In the North, killing is one method by which Inuit care for themselves and their kin. Whether this constitutes the good of killing is not mine to say. But, I will say that to kill is not always already an enactment of the bad, that killing in the North is governed by a different set of ethical and ontological axioms.

Take, for example, a short film produced by Tagaq and Zacharias Kunuk called *Tungijuq* (2010), which imagines a world in which Inuit and animals are co-constituted subjects that both hold the right to kill.[28] Tagaq and Kunuk shore up what Tina M. Campt calls "a confrontational practice of visibility," one that illuminates an ontology in which species is ungovernable, where a human–animal binary cannot survive the weather.[29] Tagaq is both human and animal, hunter and hunted, and it is because of this ontological mixing that the sights and sounds of killing become enmeshed in an affective structure we might call intimacy. This suggests that (1) the North is not governed by a logic of anthropocentrism that makes animals into always already objects of injury and (2) Inuit relations to animals and vice versa are not abstracted, that they co-produce entangled worlds where power is widely distributed. The soundscape of the film, which includes blowing snow and Inuit throat singing, keys us into an environment that skirts analytic description. Killing might not be caring, but in the North it is nothing like an exercise of anthropocentric power.

I want to end with Naisargi Dave's "the tyranny of consistency," which, in Dave's words, "is related very much to identity politics and the recognition that people who do things normatively, who eat anything, who are heterosexual, whatever the case, they do not have to explain the hypocrisies in their own life. It's only the person who tries to do something different, who is then subject to the problem of contradiction." We are asked to disavow paradox, which is to say that paradox is singled out as an impasse to political action. But worlds, especially Indigenous ones, proliferate paradox. So, if we think paradoxically, we avoid blurring the minutiae of how decisions are made by those who live in shoddy social worlds conditioned by strained agency. Paradox prevents us from holding out hope for a universal theory of ethical living. In short, paradox provincializes.

NOTES

1. Brad Wheeler, "Polaris Prize Winner Tanya Tagaq on Her Controversial Acceptance Speech," *Globe and Mail*, September 23, 2014. https://www.theglobeandmail.com/arts/awards-and-festivals/polaris-prize-winner-tanya-tagaq-on-her-controversial-acceptance-speech/article20747538/.
2. Ibid.
3. Lisa Stevenson, *Life Beside Itself: Imagining Care in the Canadian Arctic* (Berkeley: University of California Press, 2014).
4. Keavy Martin, *Stories in a New Skin: Approaches to Inuit Literature* (Winnipeg: University of Manitoba Press, 2012).
5. Wheeler, "Polaris Prize Winner."
6. Ibid.
7. Dave Dean, "Tanya Tagaq's Cute Sealfie Pissed Off a Lot of Idiots," VICE, 2014. https://www.vice.com/en_ca/article/4w7awj/tanya-taqaqs-cute-sealfie-pissed-off-a-lot-of-idiots.
8. Veena Das, *Life and Words: Violence and the Descent into the Ordinary* (Berkeley: University of California Press, 2007), 79.
9. Dean, "Tanya Tagaq's Cute Sealfie."
10. Ibid.
11. Ibid.
12. Judith Butler, "Judith Butler: Speaking of Rage and Grief." YouTube video, 11:31. Posted by Leigha Cohen, 2014. https://www.youtube.com/watch?v=ZxyabzopQi8.
13. Stevenson, *Life Beside Itself.*
14. Ibid.
15. Martin, *Stories in a New Skin.*
16. CBC News, "Inuit gather in Iqaluit for pro-sealing, anti-Ellen #sealfie." CBC, 2014. http://www.cbc.ca/news/canada/north/inuit-gather-in-iqaluit-for-pro-sealing-anti-ellen-sealfie-1.2589012.
17. Alexis Shotwell, *Against Purity: Living Ethically in Compromised Times* (Minneapolis: University of Minnesota Press, 2016), 3.
18. Ibid., 5.

19. Ibid., 8.

20. Ibid., 5.

21. For a discussion of fish pluralities in the hamlet of Paulatuuq, Northwest Territories, see Zoe S. Todd, "Fish Pluralities: Human–animal relations and sites of engagement in Paulatuuq, Artic Canada," *Études/Inuit/Studies* 38, nos. 1–2 (2012): 217–238.

22. Shotwell, *Against Purity*, 107.

23. Elizabeth Povinelli, *Economies of Abandonment: Social Belonging and Endurance in Late Liberalism* (Durham, N.C.: Duke University Press, 2011).

24. https://www.academia.edu/19629267/Ethics_as_An_Expression_of_Everyday_Life_Revised_.

25. Todd, "Fish Pluralities."

26. Donna Haraway, *Staying with the Trouble: Making Kin in the Chthulucene* (Durham, N.C.: Duke University Press, 2016), 1.

27. Joshua Ostroff, "'Angry Inuk' Explores the Inuit Fight to Protect the Seal Hunt." Huffington Post, 2016. http://www.huffingtonpost.ca/2016/11/19/angry-inuk-film_n_12527482.html.

28. See http://www.isuma.tv/tungijuq/tungijuq720p.

29. Tina M. Campt, *Listening to Images* (Durham, N.C.: Duke University Press, 2017).

ACKNOWLEDGMENTS

First and foremost, we are grateful to the participants and respondents who gave their time, energy, and thought to this project. Many other people provided crucial assistance and encouragement as the work unfolded: Our colleague Mick Smith was a generous and engaging pilot interviewee who helped us refine our focus in important ways. A dedicated team of research and technical assistants—Elysia Ackroyd, Ashley Hanna, Krista Magee, Andjela Maslenjak, Kelly Struthers Montford, Christine Moon, Erin Smiley, and Andrew Surya—recorded and transcribed interviews, conducted background research, and communicated with interviewees. Mary Louise Adams, Laurel Aziz, Jean Bruce, Eleanor MacDonald, Radhika Mongia, and Willy Vestering each provided sage advice and moral support over the long period during which this project unfolded. The APPLE (Animals in Philosophy, Politics, Law, and Ethics) group and the Lives of Animals Research Group at Queen's invited our participation in their events, asked generative questions, and helped foster a supportive and intellectually rich environment for the gestation of this work. Three reviewers—Ronald Broglio, Chad Lavin, and Anita Mannur—offered generous and valuable feedback. And Nancy Jo Cullen was a meticulous and accommodating proofreader. Finally, our thanks go to Richard Morrison, our editor, for his vision, insight, and support.

RECOMMENDED READING

Adams, Carol J. *The Sexual Politics of Meat: A Feminist-Vegetarian Critical Theory*. New York: Continuum, 1990.

Adams, Carol J., and Josephine Donovan, eds. *Animals and Women: Feminist Theoretical Explorations*. Durham, N.C.: Duke University Press, 1995.

Ahuja, Neel. *Bioinsecurities: Disease Interventions, Empire, and the Government of Species*. Durham, N.C.: Duke University Press, 2016.

———. "Postcolonial Critique in a Multispecies World." *PMLA* 124, no. 2 (2009): 556–563.

Anderson, Virginia. *Creatures of Empire: How Domestic Animals Transformed Early America*. New York: Oxford University Press, 2006.

Bekoff, Marc. *Animal Passions and Beastly Virtues: Reflections on Redecorating Nature*. Philadelphia: Temple University Press, 2005.

———. *The Emotional Lives of Animals: A Leading Scientist Explores Animal Joy, Sorrow, and Empathy—and Why They Matter*. Novato, Calif.: New World Library, 2007.

Belcourt, Billy-Ray. "Animal Bodies, Colonial Subjects: (Re)locating Animality in Decolonial Thought." *Societies* 5, no. 1 (2014): 1–11.

Bryant, Taime L. "Similarity or Difference as a Basis for Justice: Must Animals Be Like Humans to Be Legally Protected from Humans?" *Law and Contemporary Problems* 70, no. 1 (2007): 207–254.

Cairns, Kate, and Joseé Johnston. "On (Not) Knowing Where Your Food Comes From: Meat, Mothering and Ethical Eating." *Agriculture and Human Values* 35, no. 3 (2018): 569–580.

Calarco, Matthew. "Being-Toward-Meat: Anthropocentrism, Indistinction, and Veganism." *Dialectical Anthropology* 38 (2014): 415–429.

———. "Deconstruction Is Not Vegetarianism: Humanism, Subjectivity, and Animal Ethics." *Continental Philosophy Review* 37, no. 2 (2004): 175–201.

———. "Identity, Difference, Indistinction." *CR: The New Centennial Review* 11, no. 2 (2011): 41–60.

———. "We Are Made of Meat," An Interview by Leonardo Caffo with Matthew Calarco." *Animal Rights Zone*. June 12, 2013. http://arzone.ning.com/profiles/blogs/we-are-made-of-meat-the-matthew-calarco-interview.

Cavalieri, Paola, ed. *The Death of the Animal: A Dialogue*. New York: Columbia University Press, 2009.
Chen, Mel Y. *Animacies: Biopolitics, Racial Mattering, and Queer Affect*. Durham, N.C.: Duke University Press, 2012.
Chrulew, Matthew. "Animals in Bio-Political Theory: Between Agamben and Negri." *New Formations* 76, (2012): 53–67.
Cixous, Hélène. "From My Menagerie to Philosophy." In *Resistance, Flight, Creation: Feminist Enactments of French Philosophy*, ed. Dorothea Olkowski, 40–47. Ithaca, N.Y.: Cornell University Press, 2000.
Coetzee, John M. *Elizabeth Costello*. New York: Viking, 2003.
———. *The Lives of Animals*. Princeton, N.J.: Princeton University Press, 1999.
Corman, Lauren. "Capitalism, Veganism, and the Animal Industrial Complex." *Rabble* (Blog). April 21, 2014. http://rabble.ca/blogs/bloggers/vegan-challenge/2014/04/capitalism-veganism-and-animal-industrial-complex.
———. "Getting Their Hands Dirty: Racoons, Freegans, and Urban 'Trash.'" *Journal for Critical Animal Studies* IX, no. 3 (2011): 28–61.
———. "The Ventriloquist's Burden? Animals, Voice, and Politics." Doctoral dissertation, York University, 2012.
Corman, Lauren, and Teresa Vandrovcová. "Radical Humility: Towards a More Holistic Critical Animal Studies Pedagogy." In *Defining Critical Animal Studies: An Intersectional, Social Justice Approach for Total Liberation*, ed. Anthony J. Nocella II, John Sorenson, Kim Socha, and Atsuko Matsuoka, 135–157. Bern: Peter Lang, 2014.
Dalal, Neil, and Chloë Taylor. *Asian Perspectives on Animal Ethics: Rethinking the Nonhuman*. New York: Routledge, 2014.
Dave, Naisargi N. "Something, Everything, Nothing; Or, Cows, Dogs, and Maggots." *Social Text* 35, no. 1.130 (2017): 37–57. doi: 10.1215/01642472-3727984.
———. "Witness: Humans, Animals, and the Politics of Becoming." *Cultural Anthropology* 29, no. 3 (2014): 433–456. doi:10.14506/ca29.3.01.
Davis, Karen. "The Mental Life of Chickens as Observed Through Their Social Relationships." In *Experiencing Animal Minds: An Anthology of Animal–Human Encounters*, ed. Julie A. Smith and Robert W. Mitchell, 13–29. New York: Columbia University Press, 2012.
Deckha, Maneesha. "Milk's Global Rise: A Case Study to Illuminate the Transspecies Violence of Law and Colonialism." Review of "Animal Colonialism: The Case of Milk," by Mathilde Cohen, *American Journal of International Law Unbound* 267 (2017). *Jotwell*, April 25, 2018. https://equality.jotwell.com/milks-global-rise-a-case-study-to-illuminate-the-transspecies-violence-of-law-and-colonialism/.

Derrida, Jacques. *The Animal That Therefore I Am*. New York: Fordham University Press, 2008.

Derrida, Jacques, and Jean-Luc Nancy. "'Eating Well,' or the Calculation of the Subject: An Interview with Jacques Derrida." In *Who Comes After the Subject?*, ed. Eduardo Cadava et al., 96–119. New York: Routledge, 1991.

Despret, Vinciane. *What Would Animals Say If We Asked the Right Questions?*, trans. Brett Buchanan. Minneapolis: University of Minnesota Press, 2016.

De Waal, Frans. *The Bonobo and the Atheist: In Search of Humanism Among the Primates*. New York: Norton, 2013.

Donaldson, Brianne. *Creaturely Cosmologies: Why Metaphysics Matters for Animal and Planetary Liberation*. London: Lexington Books, 2015.

Donaldson, Sue, and Will Kymlicka. *Zoopolis: A Political Theory of Animal Rights*. Oxford: Oxford University Press, 2011.

DuPuis, E. Melania, and David Goodman. "Should We Go 'Home' to Eat? Toward a Reflexive Politics of Localism." *Journal of Rural Studies* 21, no. 3 (2005): 359–371.

Emel, Jody, and Harvey Neo, eds. *Political Ecologies of Meat*. London: Routledge, 2015.

Francione, Gary, and Anna E. Charlton. "Veganism without Animal Rights." *The European* (Website), July 13, 2015. https://www.theeuropean-magazine.com/gary-l-francione/10366–the-morality-of-eating-meat-eggs-and-dairy.

Gaard, Greta. "Towards a Feminist Postcolonial Milk Studies." *American Quarterly* 65, no. 3 (2013): 595–618.

García, María Elena. "Grieving Guinea Pigs: Reflections on Research and Shame in Peru." In *Grieving Witnesses: The Politics of Grief in the Field*, ed. Kathryn Gillespie and Patricia Lopez. Berkeley: University of California Press, forthcoming.

———. "How Guinea Pigs Work: Figurations and GastroPolitics in Peru." In *How Nature Works*, ed. Alexander Blanchette, Sarah Besky, and Naisargi Dave. Albuquerque: University of New Mexico Press, 2010.

———. "Love, Death, Food, and Other Ghost Stories: Hauntings of Intimacy and Violence in Contemporary Peru." In *Economies of Death: Economic Logics of Killable Life and Grievable Death*, ed. Kathryn Gillespie and Patricia Lopez, 160–178. New York: Routledge, 2015.

———. "Super Guinea Pigs?" *Anthropology Now* 2, no. 2 (forthcoming 2019).

———. "The Taste of Conquest: Colonialism, Cosmopolitics, and the Dark Side of Peru's Gastronomic Boom." *Journal of Latin American and Caribbean Anthropology* 18, no. 3 (2013): 505–524.

Gossett, Che. "Blackness, Animality, and the Unsovereign." *Verso*. September 15, 2015. https://www.versobooks.com/blogs/2228-che-gossett-blackness-animality-and-the-unsovereign.

Graeber, David. "What's the Point If We Can't Have Fun?" *The Baffler*, no. 24 (2014). https://thebaffler.com/salvos/whats-the-point-if-we-cant-have-fun.
Gruen, Lori, and Kari Weil. "Animal Others—Editors' Introduction." *Hypatia* 27, no. 3 (2012), 477–487.
Guthman, Julie. "Fast Food/Organic Food: Reflexive Tastes and the Making of 'Yuppie Chow.'" *Social and Cultural Geography* 4, no. 1 (2003): 45–58.
Haraway, Donna J. *The Companion Species Manifesto: Dogs, People, and Significant Otherness*. Chicago: Prickly Paradigm Press, 2003.
———. *When Species Meet*. Minneapolis: University of Minnesota Press, 2008.
Haraway, Donna J., and Cary Wolfe. *Manifestly Haraway*. Minneapolis: University of Minnesota Press, 2016.
Harper, A. Breeze, ed. *Sistah Vegan: Black Female Vegans Speak on Food, Identity, Health, and Society*. Brooklyn: Lantern Books, 2010.
Hearne, Vicki. *Adam's Task: Calling Animals by Name*. New York: Knopf, 1986.
Holland, Sharon P. "(Black) (Queer) Love." *Callalo* 36, no. 3 (2013): 658–668.
———. *The Erotic Life of Racism*. Durham, N.C.: Duke University Press, 2012.
———. *the professor's table* (Blog). https://theprofessorstable.wordpress.com/.
———. *Raising the Dead: Readings of Death and (Black) Subjectivity*. Durham, N.C.: Duke University Press, 2000.
Horvoka, Alice. Foreword to *Political Ecologies of Meat*, ed. Jody Emel and Harvey Neo, 1–18. London: Routledge, 2015.
Hribal, Jason. *Fear of the Animal Planet: The Hidden History of Animal Resistance*. Chico, Calif.: AK Press, 2011.
Hua, Juliette, and Neel Ahuja. "Chimpanzee Sanctuary: 'Surplus' Life and the Politics of Transspecies Care." *American Quarterly* 65, no. 3 (2013): 619–637.
Isfahani-Hammond, Alexander. "Of She-Wolves and Mad Cows: Animality, Anthropophagy and the State of Exception in Cláudio Assis's 'Amerelo Mango." *Luso-Brazilian Review* 48, no. 2 (2011): 129–149.
Jackson, Zakiyyah. "Animal: New Directions in the Theorization of Race and Posthumanism." *Feminist Studies* 39, no. 3. (2013): 669–685.
Jenkins, Stephanie, and Vasile Stanescu. "One Struggle." In *Defining Critical Animal Studies: An Intersectional Social Justice Approach to Animal Liberation*, ed. Anthony J. Nocella, John Sorensen, Kim Socha, and Atsuko Matsuoka, 74–88. New York: Peter Lang, 2014.
Johnston, Josée, Michelle Szabo, and Alexandra Rodney. "Good Food, Good People: Understanding the Cultural Repertoire of Ethical Eating." *Journal of Consumer Culture* 11, no. 3 (2011): 293–318.
Kheel, Marti. *Nature Ethics: An Ecofeminist Perspective*. Plymouth: Rowman & Littlefield, 2008.

Kim, Claire Jean. *Dangerous Crossings: Race, Species, and Nature in a Multicultural Age*. Cambridge: Cambridge University Press, 2015.
Kim, Claire Jean, and Carla Freccero. "Introduction: A Dialogue." *American Quarterly* 65, no. 3 (2013): 461.
Kingsolver, Barbara. *Animal, Vegetable, Miracle: A Year of Food Life*, first edition. New York: HarperCollins, 2007.
Lavin, Chad. *Eating Anxiety: The Perils of Food Politics*. Minneapolis: University of Minnesota Press, 2013.
Livingston, Julie, and J. Puar. "Interspecies." *Social Text* 29, no. 1 (2011): 3–14.
Mallory, Chaone. "Locating Ecofeminism in Encounters with Food and Place." *Journal of Agricultural and Environmental Ethics* 26, no. 1 (2013): 171–189.
Mannur, Anita. "Food Matters: An Introduction," *Massachusetts Review* 45, no. 3 (2004): 209–215.
McCance, Dawn. *Critical Animal Studies: An Introduction*. New York: SUNY Press, 2013.
McHugh, Susan. "'A Flash Point in Inuit Memories': Endangered Knowledges in the Mountie Sled Dog Massacre." *English Studies in Canada* 39, no. 1 (2013): 149–175.
Moore, Jason. *Capitalism in the Web of Life: Ecology and the Accumulation of Capital*. London: Verso, 2015.
Otsuki, Grant Jun. "Naisargi Dave on Animal Rights in India." AnthroPod: The SCA Podcast, *Cultural Anthropology* (Website), February 2, 2015. https://culanth.org/fieldsights/629–naisargi-dave-on-animal-rights-activism-in-india.
Nadasdy, Paul. *Hunters and Bureaucrats: Power, Knowledge, and the Aboriginal-State Relations in the Southwest Yukon*. Vancouver: University of British Columbia Press, 2003.
Nibert, David A. *Animal Oppression and Human Violence: Domesecration, Capitalism and Human Conflict*. New York: Columbia University Press, 2013.
Nocella, Anthony J., II, John Sorenson, Kim Socha, and Atsuko Matsuoka. "The Emergence of Critical Animal Studies: The Rise of Intersectional Animal Liberation," Introduction to *Defining Critical Animal Studies: An Intersectional Social Justice Approach for Liberation*, ed. Anthony J. Nocella II, John Sorenson, Kim Socha, and Atsuko Matsuoka, xix–xxxvi. New York: Peter Lang, 2014.
Pachirat, Timothy. *Every Twelve Seconds: Industrialized Slaughter and the Politics of Sight*. New Haven, Conn.: Yale University Press, 2011.
Parreñas, Juno (Rheana) Salazar. *Decolonizing Extinction: The Work of Care in Orangutan Rehabilitation*. Durham, N.C.: Duke University Press, 2018.

———. *Gender: Animals*. Macmillan Interdisciplinary Handbooks. Farmington Hills, Mich.: Macmillan Reference USA, 2017.

———. "The Materiality of Intimacy in Wildlife Rehabilitation: Rethinking 'Ethical Capitalism' through Embodied Encounters with Animals in Southeast Asia." *Positions* 24, no. 1 (2016): 97–127.

Pilgrim, Karyn. "'Happy Cows,' 'Happy Beef': A Critique of the Rationales for Ethical Meat." *Environmental Humanities* 3, no. 1 (2013): 111–127.

Plumwood, Val. "Being Prey." *Utne Reader* (2000): 56–61.

Pollan, Michael. *The Omnivore's Dilemma: A Natural History of Four Meals*. New York: Penguin, 2006.

Potter, Will. *Green Is the New Red: An Insider's Account of a Social Movement Under Siege*. San Francisco: City Lights, 2011.

Potts, Annie, ed. *Meat Culture*. Leiden, Netherlands: Brill, 2016.

Puig de la Bellacasa, Maria. *Matters of Care: Speculative Ethics in More Than Human Worlds*. Minneapolis: University of Minnesota Press, 2017.

Rader, Karen. *Making Mice: Standardizing Animals for American Biomedical Research, 1900–1955*. Princeton, N.J.: Princeton University Press, 2004.

Regan, Tom. *The Case for Animal Rights*, first edition. Berkeley: University of California Press, 2004.

Robinson, Margaret. "Veganism and Mi'kmaq Legends." *Canadian Journal of Native Studies* 33, no. 1 (2013): 189–96.

Safran Foer, Jonathan. *Eating Animals*. London: Penguin, 2009.

Sassatelli, Roberta, and Davolio, Federica. "Consumption, Pleasure, and Politics: Slow Food and the Politico-Aesthetic Problematization of Food." *Journal of Consumer Culture* 10, no. 2 (2010): 202–232.

Shotwell, Alexis. *Against Purity: Living Ethically in Compromised Times*. Minneapolis: University of Minnesota Press, 2016. E-book.

Shukin, Nicole. *Animal Capital: Rendering Life in Biopolitical Times*. Minneapolis: University of Minnesota Press, 2009.

Singer, Peter. *Animal Liberation: A New Ethics for Our Treatment of Animals*. New York: Avon, 1975.

Smuts, Barbara. "Between Species: Science and Subjectivity." *Configurations* 14, nos. 1–2 (2006): 115–126.

Singh, Bhrigupati, and Naisargi Dave. "On the Killing and Killability of Animals: Nonmoral Thoughts for the Anthropology of Ethics." *Comparative Studies of South Asia, Africa and the Middle East* 35, no.2 (2015): 232–245. doi: 10.1215/1089201x-3139012.

Smith, Mick. *Against Ecological Sovereignty: Ethics, Biopolitics, and Saving the Natural World*. Minneapolis: University of Minnesota Press, 2011.

Sorensen, John. *About Canada: Animal Rights*. Halifax, N.S.: Fernwood Press, 2010.

———. *Constructing Ecoterrorism: Capitalism, Speciesism and Animal Rights.* Halifax, N.S.: Fernwood Press, 2016.

Stanescu, James. "Beyond Biopolitics: Animal Studies, Factory Farms, and the Advent of Deading Life." *PhaenEx* 8, no. 2 (2013): 135–160.

Stanescu, Vasile. "'Green' Eggs and Ham? The Myth of Sustainable Meat and the Danger of the Local." *Journal of Critical Animal Studies* 1, no. 11 (2010): 8–32.

Steeves, H. Peter. "And Say the Zombie Responded? or, How I Learned to Stop Living and Unlove the Undead." In *Beautiful, Bright, and Blinding: Phenomenological Aesthetics and the Life of Art*, 113–138. Albany: SUNY Press, 2017.

———. "The Boundaries of the Phenomenological Community: Non-Human Life and the Extent of our Moral Enmeshment." In *Becoming Persons*, ed. Robert N. Fisher, 777–797. Oxford: Applied Theology Press, 1995.

———. "The Man Who Mistook His Meal for a Hot Dog." In *Beautiful, Bright, and Blinding: Phenomenological Aesthetics and the Life of Art*, 139–156. Albany: SUNY Press, 2017.

———. "They Say Animals Can Smell Fear." In *Animal Others: On Ethics, Ontology, and Animal Life*, ed. H. Peter Steeves, 133–178. Albany: SUNY Press, 1999.

Steinbock, Eliza, Marianna Szczygielska, and Anthony Wagner. *Thinking Linking*. Editorial Introduction to *Tranimacies: Intimate Links Between Animal and Trans* Studies*. Special Issue of *Angelaki: Journal of Theoretical Humanities* 22, no. 2 (2017): 1–10.

Steinfeld, Henning, et al. "Livestock's Long Shadow." *Rome: Food and Agriculture Organization of the United Nations* (2006), http://www.fao.org/docrep/010/a0701e/a0701e00.HTM.

Struthers Montford, Kelly. "Agricultural Power: Politicized Ontologies of Food, Life, and Law in Settler Colonial Spaces." Doctoral dissertation, University of Alberta, 2017.

TallBear, Kim. "Why Interspecies Thinking Needs Indigenous Standpoints." Presentation at the Annual Meeting of the American Anthropological Association, Montreal, QC, November 16–20, 2011.

Taylor, Chloë. "Abnormal Appetites: Foucault, Atwood, and the Normalization of a Meat-Based Diet." *Journal for Critical Animal Studies* 10, no. 4 (2012): 130–148.

Taylor, Sunaura. *Beasts of Burden: Animal and Disability Liberation*. New York: The New Press, 2017.

Thomas, Elizabeth Marshall. *The Hidden Life of Dogs*. New York: Mariner Books, 2010.

Todd, Zoe S. "Fish Pluralities: Human–Animal Relations and Sites of Engagement in Paulatuuq, Arctic Canada." *Études/Inuit/Studies* 38, no. 1–2 (2012): 217–238.

Twine, Richard. "Vegan Killjoys at the Table: Contesting Happiness and Negotiating Relationships with Food Practices." *Societies* 4, no. 4 (2014): 623–629.

Wadiwel, Dinesh. *The War Against Animals*. Leiden, Netherlands: Brill, 2015.

Weaver, Harlan. "'Becoming in Kind': Race, Class, Gender, and Nation in Cultures of Dog Rescue and Dog Fighting." *American Quarterly*, 65, no. 3 (2013): 689–709.

———. "Pit Bull Promises: Inhuman Intimacies and Queer Kinships in an Animal Shelter." *GLQ* 21, no. 2–3 (2015): 343–363.

———. "Trans Species." *Trans*gender Studies Quarterly* 1, no. 102 (2014): 253–254.

Weheliye, Alexander. *Habeus Viscus: Racializing Assemblages, Biopolitics, and Black Feminist Theories of the Human.* Durham, N.C.: Duke University Press, 2014.

Weil, Kari. "They Eat Horses, Don't They? Hippophagy and Frenchness." *Gastronomica: The Journal of Critical Food Studies* 7, no. 2 (2007): 44–51. doi: 10.1525/gfc.2007.7.2.44.

———. *Thinking Animals: Why Animal Studies Now*? New York: Columbia University Press, 2012.

Weis, Tony. "The Meat of the Global Food Crisis." *Journal of Peasant Studies* 40, no. 1 (2013b): 65–68.

Weisberg, Zipporah. "The Broken Promises of Monsters: Haraway, Animals, and the Humanist Legacy." *Journal for Critical Animal Studies* 7, no. 2 (2009): 22–62.

Weiss, Erica. "'There Are No Chickens in Suicide Vests': The Decoupling of Human Rights and Animal Rights in Israel." *Journal of the Royal Anthropological Institute* 22, no. 3 (2016). doi:10.1111/1467–9655.12453.

Wheeler, Andrea. "An Interview with Luce Irigaray on *Vegetal Being: Two Philosophical Perspectives* and Sexuate Difference." *Angelaki: Journal of the Theoretical Humanities* 22, no. 4 (2017): 177–181.

Wolfe, Cary. *Animal Rites: American Culture, The Discourse of Species, and Posthumanist Theory*. Chicago: University of Chicago Press, 2003.

———. *Before the Law: Humans and Other Animals in a Biopolitical Frame*. Chicago: University of Chicago Press, 2012.

———. *What Is Posthumanism?* Minneapolis: University of Minnesota Press, 2010.

Wolfe, Cary, ed. *Zoontologies: The Question of the Animal*. Minneapolis: University of Minnesota Press, 2003.
Wood, David. "Comment ne pas Manger—Deconstruction and Humanism." In *Animal Others: On Ethics, Ontology and Animal Life*, ed. H. Peter Steeves, 12–25. Albany: SUNY Press, 1999.
Wright, Laura. *The Vegan Studies Project: Food, Animals, and Gender in the Age of Terror*. Athens: University of Georgia Press, 2015.

CONTRIBUTORS

NEEL AHUJA is Associate Professor in Feminist Studies and Critical Race and Ethnic Studies at the University of California, Santa Cruz. He is the author of *Bioinsecurities: Disease Interventions, Empire, and the Government of the Species* (Duke University Press, 2016). His articles have appeared in *GLQ*, *Social Text*, and *PMLA*, among other venues.

BILLY-RAY BELCOURT is a scholar and poet from the Driftpile Cree Nation. A Rhodes Scholar at the University of Oxford from 2016–17, he is currently a PhD student at the University of Alberta. Belcourt's work has appeared in *Settler Colonial Studies* and *Societies*, among other venues. He is the author of the poetry collection *This Wound Is a World*, which won the 2018 Griffin Poetry Prize.

MATTHEW R. CALARCO is Professor of Philosophy at California State University, Fullerton. He is the author of *Zoographies: The Question of the Animal from Heidegger to Derrida* (Columbia University Press, 2008) and *Thinking Through Animals: Identity, Difference, Indistinction* (Stanford University Press, 2015).

R. SCOTT CAREY is a grant writer in Calgary, Alberta. He has a PhD in Kinesiology and Health Studies from Queen's University and has published his research in *Sport in Society* and *Body & Society*, among other venues.

LAUREN CORMAN is Associate Professor in the Department of Sociology at Brock University. She is the editor of a special issue of *UnderCurrents: Journal of Critical Environmental Studies*, co-editor of *Animal Subjects 2.0* (Wilfrid Laurier University Press, 2016), and the founder and former producer of the radio program *Animal Voices*.

NAISARGI N. DAVE is Associate Professor of Anthropology at the University of Toronto and author of *Queer Activism in India: A Story in the Anthropology of Ethics* (Duke University Press, 2012). Her work has appeared in

Social Text, *Cultural Anthropology*, and *Feminist Studies*, among other venues. Her book in progress is titled *The Social Skin: Humans and Animals in India*.

Maneesha Deckha is Professor and the Lansdowne Chair in Law at the University of Victoria. She is the author of numerous articles published in such journals as *Ethics & the Environment*, *Journal of Animal Law and Ethics*, *Hypatia*, the *McGill Law Journal*, and the *Harvard Journal of Law & Gender*. She has also contributed to a variety of edited collections in feminist legal, postcolonial, and critical animal studies.

María Elena García is Associate Professor in the Comparative History of Ideas program and the Jackson School of International Studies at the University of Washington. She is the author of *Making Indigenous Citizens: Identities, Development, and Multicultural Activism in Peru* (Stanford University Press, 2005), with a second book, *Culinary Spectacles: Gastro-Politics and Other Tales of Race and Species in Peru*, under contract with the University of California Press.

Sharon P. Holland is the Townsend Ludington Term Distinguished Endowed Professor in the Department of American Studies at the University of North Carolina at Chapel Hill. She is author of *Raising the Dead: Readings of Death and (Black) Subjectivity* (Duke University Press, 2000) and *The Erotic Life of Racism* (Duke University Press, 2012). She blogs at http://theprofessorstable.wordpress.com/ and is currently working on an investigation of the human–animal distinction, and the place of discourse on blackness within that discussion.

Samantha King is Professor of Gender Studies and Kinesiology and Health Studies at Queen's University. Her work has appeared in *Social Text*, *Ethnic and Racial Studies*, *Health Communication*, and the *International Journal of Drug Policy*, among other venues. Her book *Pink Ribbons, Inc: Breast Cancer and the Politics of Philanthropy* (University of Minnesota Press, 2006) is the subject of a National Film Board documentary by the same name.

Isabel Macquarrie is a Juris Doctor candidate at Harvard Law School. Prior to law school, Isabel completed a Master of Arts in sociology at Queen's University, where her research focused on programming for disadvantaged youth. After law school Isabel plans to work in legal aid.

Victoria N. Millious is a PhD candidate in the School of Kinesiology and Health Studies, Queen's University. Her doctoral project theorizes the phenomena of perceived insufficient milk (PIM) in contemporary breastfeeding/chestfeeding practices from a feminist and anti-healthist orientation. When she has completed her PhD, she will continue working with nonprofit organizations that address gender and health-related inequalities.

Elaine M. Power is Associate Professor in the School of Kinesiology & Health Studies at Queen's University. She is co-author of *Acquired Tastes: Why Families Eat the Way They Do* (University of British Columbia Press, 2015), co-editor of *Neoliberal Governance and Health: Duties, Risks and Vulnerabilities* (McGill-Queen's University Press, 2016), and co-editor of the forthcoming volume *Feminist Food Studies: Intersectional Perspectives* (Women's Press). She is a founding member of the Canadian Association for Food Studies (CAFS).

H. Peter Steeves is Professor of Philosophy and Director of the Humanities Center at DePaul University. He is the author of several books, including *Animal Others: On Ethics, Ontology, and Animal Life* (SUNY Press, 1999), *The Things Themselves: Phenomenology and the Return to the Everyday* (SUNY Press, 2006), and *Beautiful, Bright, and Blinding: Phenomenological Aesthetics and the Life of Art* (SUNY Press, 2017).

Kelly Struthers Montford is a Postdoctoral Research Fellow in Punishment, Law and Social Theory at the University of Toronto. She received her PhD in sociology from the University of Alberta in 2017. Her work has appeared in the *New Criminal Law Review*, *PhiloSophia*, and the *Canadian Journal of Women and the Law*, among other venues.

Kim TallBear is Associate Professor and Canada Research Chair in Indigenous Peoples, Technoscience, and Environment in the Faculty of Native Studies at the University of Alberta. She is author of *Native American DNA: Tribal Belonging and the False Promise of Genetic Science* (University of Minnesota Press, 2013) and a member of the Oak Lake Writers, a group of Dakota, Lakota, and Nakota (Oceti Sakowin) writers.

Sunaura Taylor is an artist and writer based in New York City and the author of *Beasts of Burden: Animal and Disability Liberation* (The New Press, 2017). She has written for *AlterNet*, *American Quarterly*, *BOMB*, the *Monthly Review*, *Qui Parle*, and *Yes!* magazine and has contributed to the books *Ecofeminism*, *Defiant Daughters*, *Occupy!*; *Stay Solid*; and *Infinite City*.

Harlan Weaver is Assistant Professor in the Gender, Women, and Sexuality Studies Department at Kansas State University. He has written articles on queer intimacies in multispecies ethnography; interspecies intersectionalities in dog fighting, rescue, and training; and trans affect in *GLQ*, *Catalyst*, *American Quarterly*, and *Somatechnics*.

Kari Weil is University Professor of Letters at Wesleyan University. She is the co-editor of a special issue of *Hypatia* entitled "Animal Others" (Gruen & Weil, 2012) and author of *Thinking Animals: Why Animal Studies Now* (Columbia University Press, 2012). Her book *Horses and Their Humans in Nineteenth-Century France: Mobility, Magnetism, Meat* is forthcoming.

Cary Wolfe is the Bruce and Elizabeth Dunlevie Professor of English at Rice University. He is author of *Animal Rites: American Culture, the Discourse of Species*, and *Posthumanist Theory* (University of Chicago Press, 2003) and *Before the Law: Humans and Other Animals in a Biopolitical Frame* (University of Chicago Press, 2012).

INDEX